D0389116

BLADE DANCER

ALSO BY S. L. VIEHL

StarDoc
Beyond Varallan
Endurance
Shockball
Eternity Row

BLADE DANCER

S. L. VIEHL

A ROC BOOK

ROC
Published by New American Library, a division of
Penguin Group (USA) Inc., 375 Hudson Street,
New York, New York 10014, U.S.A.
Penguin Books Ltd, 80 Strand,
London WC2R 0RL, England
Penguin Books Australia Ltd, 250 Camberwell Road,
Camberwell, Victoria 3124, Australia
Penguin Books Canada Ltd, 10 Alcorn Avenue,
Toronto, Ontario, Canada M4V 3B2
Penguin Books (N.Z.) Ltd, Cnr Rosedale and Airborne Roads,
Albany, Auckland 1310, New Zealand

Penguin Books Ltd, Registered Offices:
80 Strand, London WC2R 0RL, England

First published by Roc, an imprint of New American Library,
a division of Penguin Group (USA) Inc.

ISBN 0-451-45926-1

Printed in the United States of America
Set in Sabon
Designed by Ginger Legato

For my mother, Joan Jean Sabella,
who taught me what it means to have
strength, stamina, and faith.
Love you, Mom.

CHAPTER ONE

"THE PATH CHANGES, SO TOO MUST THE TRAVELER."

—TAREK VARENA, CLANJOREN

All I was trying to do when they caught me was bury my mother in an unmarked grave.

I should have seen them coming—I was out there alone, in the desert, in the middle of the night—but I didn't. Maybe because I was tired. And upset. And committing a felony.

The tired part came from staying awake since getting Mom's last signal. I'd ditched practice, dodged my offcoach, disguised myself to avoid any stray media hounds, rented a glidecar under an alias, then headed from the city for the desert. All that took forty-nine hours, with no time for a catnap in between.

I hadn't gotten upset until I'd walked through the front door of our house.

I tried not to think about that as I carried my mother's body over a mile out to the south ridge. I knew it would be safer to burn her, but I couldn't bring myself to get the petrol and the matches. I couldn't do that, not to her. I'd already done enough.

Like the felony part.

Have to finish this before dawn or someone'll spot me.

On the way to where I figured no one would ever find her body, I had to deal with the uneven terrain on top of my exhaustion. If I fell, I might end up cracking my skull, or worse, junking my knee. In the distance I saw buzzards circling over something else dead—which reminded me, I'd have to make the hole deep or they'd get at her. Other things waited in the shadows and watched for a chance at a midnight snack, too. Vultures. Coyotes. Rodents. I'd once seen a dead ewe out here covered with a seething blanket of black roaches.

I won't let them chew on you, Mom.

The temperature, which had been scorching when I'd arrived, had dropped enough to make my face feel numb and stiff. Some of the local cacti flowered at night, and I could smell their thin, desperate perfume. The sweat collecting on my scalp and under my arms had no scent, but made me itch, probably because I couldn't scratch. I hated being sweaty, even when there was no one around to notice my body odor—or lack thereof.

I knew what my mother would have said right then: *You should have showered before you left the house.*

"Yeah, well, I was busy."

Her face bounced against my chest with every step I took, and left wet marks on my shirt. I was glad it was dark. I didn't want to look at her. I didn't want my last memory to be of her like this.

My last memory was bad enough.

The desert silence, something I usually enjoyed, started to get to me. It was too quiet. There should have been coyotes yipping, wind whistling, crickets chirping, *something*. Instead, all I could hear were my own footsteps thudding against the sun-baked earth, crushing weeds, skidding a little on pebbles here and there.

"Help me out here, Mom." I shifted her weight in my arms. "Haunt me or something."

An image of her sitting calmly at our kitchen table with two teacups snapped into my mind—the way she looked up as she poured. Her hand on the teakettle. The smile that had always been a little sad.

What say you tell me about the game, Jory?

That had been six months ago, but I played along and repeated what I'd told her then. "Damn Gliders gave our defense a real pounding. We tied it up right before the one-minute warning, and the whole game came down to the final play. Thought the offcoach was going to strangle the defcoach." I grinned. "Dees kicked from the forty. Linemen everywhere, dogging anyone stupid enough to run a standard pattern and just mowing 'em down."

You are not stupid, my ClanDaughter.

I ignored the ghost voice in my head and babbled on. "Only Coach and I worked out this hook play—center run for ten, loop back to left field for five, then the old dip and dive past Dees. I snagged the sphere, cradled it in, and ran. By the time those tacks realized what we'd done, nobody could get near me. Felt like I was skipping on air all the way to the zone."

You are not skipping now.

No, I wasn't. Every step made my knee click, and serious pain pulsed up through my thigh muscle, courtesy of my artificial-joint tech. I'd have to spend a couple of hours fiddling with the ligament mounts again.

I warned you not to play so hard.

"I don't play hard. I get hit hard."

Rijor would not agree with you.

That reminded me. "Did I tell you about Rij's sib? She came downside to see me. Intergal Shockball heard our junta erased Rij's name from all the Terran databases for being part fish, so the nonhuman league is going to add him to theirs and retire his old number on his homeworld. Put him in their hall of fame and everything. You'd have been tickled."

He would have made you an excellent bondmate.

He might have, if a fan hadn't noticed water leaking from a tear in Rij's uniform after a particularly rough game two years ago. That same afternoon, an angry mob had dragged him from the arena and beaten him to death.

You should have Chosen him.

"Shut up, Mom."

The only choice I needed to make was where to plant her. After looking around, I picked a natural depression in the ground unpopulated by sagebrush and weeds, and put down the body. As I straightened, the moon came out from the clouds and made her dead eyes gleam. They were all white, no pupils, no irises. Anyone who'd have seen them would have thought she was blind.

No, Jory. Mom's ghost sounded as tired as I felt. *They'd see the color of my skin, and call the authorities.*

I didn't feel sorry for her. "Not a problem anymore, is it?"

I pulled the shovel out of my backpack and put it to use. The desert ground was dry but hard-packed, and it took a few minutes to find the right angle. Even then, I had trouble getting into the rhythm of stab-push-heave. Not like I'd had a whole lot of practice digging graves in the desert in the middle of the goddamn night.

I was meant for the embrace of the stars.

Her ghost was really starting to tick me off. "Next time die on another planet, okay?"

I should have been at a memorial center in the city, giving her a decent funeral. Flowers. Church music. Discreetly anxious attendants hovering during the services until it was time to reduce her remains to sanitary, scatterable ash. But not even a bribe of World Game tickets would keep the morticians from turning us in.

Mom and I weren't even supposed to be on Terra without special short-term visas. Instead we'd resided here illegally for twenty-four years. My mother had always been in hiding, first with the other aliens in the underground tunnels beneath the city, then by becoming a complete recluse once I had earned enough to buy the desert house.

I would have done anything for you.

I jumped—that time, she'd sounded like she'd been standing right next to me—then I shoveled faster. "Couple of fans cornered me last week outside Toronto arena. They had a new baby with them, told me they'd named her Jory." I had to stop to wipe the sweat from my eyes. "Nice people, but shit, Mom, that kid looked just like a monkey."

You should have children of your own.

"Me. A mother." I snorted. "When swine become airborne, maybe."

At last I judged the grave to be deep enough, and climbed out. I'd brought a length of old linen to wrap around Mom's body. She wouldn't have liked this part, either. According to her, bodies should be wrapped in a shroud woven by the family during the "Yay-You're-Dead" party. Where she came from, they *loved* funerals.

Death is not the end of the journey. It is only a new path taken by the traveler.

"No ship, no stars, not even a grass shroud, Mom. A hole in the dirt and cotton's the best I can do."

I folded her six-fingered hands over her sunken chest. She had beautiful hands, strong and graceful and competent. I'd never seen her claws emerge, not once in twenty-four years. The numbness inside me contracted into something else. Something tight and hot and furious.

I was *burying* my mother.

Do not grieve for me, ClanDaughter.

I didn't want to grieve.

I wanted to hit something.

Like *her*.

"Why the hell didn't you call me?" The words exploded out of me. "I'd have stomped over anything to get to you! We could have gone back underground; I could have gotten the medicine in time. I could have saved you. What were you thinking?"

It was my choice. My path.

"You and your stupid fucking paths!" I kicked the shovel, sending it flying. Then I was on my knees, my arms around my abdomen, doubled over. Losing her hurt worse than anything that had ever been done to me. "How could you? How could you leave me like this? You're all I've got. All I've ever had."

Still, white-within-white eyes stared up at the stars I couldn't give her.

I honor you, Jory.

I stopped acting like a jerk, and carefully arranged the coils and braids of her black hair around her face. Most of the pustules had broken before she'd died, and a few still oozed green fluid when I touched her—the same fluid that was all over the front of my shirt. Trickles of it ran down her cheeks, like dark tears.

"At least one of us can cry." I sat back on my haunches and

pressed my palms to the sides of my pounding head. "So what do I do now, Mom?"

My answer came immediately, when light flashed in my face. "PRC. Hold it right there."

Five men surrounded me, and I curled my glove around the hilt of the knife I always carried, even during games. No one was going to do me the way they'd done Rij. The high-intensity emitters they carried made it hard to see their faces, but they were obviously well dressed. Every one of them had a weapon drawn.

With Mom dead, there'd be no more bribes from me. Evidently the neighbors had decided to go elsewhere to get some creds.

I straightened to my full height, and two of the men took an automatic step back. Nice thing about being nearly seven feet tall—it unnerved every guy I met. "Get lost."

While they were busy gaping, I picked up my mother and carried her over to the hole. They followed, forming a loose ring around me, Mom, and the grave.

"What are you burying?" one of them asked.

I could have lied and said a really big dog or something. But the reason to do that was going in the grave. What was the point? "My mother."

"Why didn't you take her to a mortuary memorial center?" The PRC agent pointed a beam at my face.

I stepped out of the light. "Because, stupid, she didn't like them. Take a hike."

"Put down the body."

I carried her to the grave and jumped in. New pain sizzled up my thigh as I laid Mom out in the bottom. The old fears came crawling along with it. Maybe I could cover her fast, make some excuses. I'd listed "future-ager" as my religious preference with the junta; that might get me some slack. Five faces stared over the edge at me. Someone enabled a weapon.

Ta-ta, slack.

"You have ten seconds to climb out of there."

I used seven to bend down and kiss my mother's ruined face. Her brow felt hard and cold against my lips. "Honor you, Mom."

I ignored the outstretched hands and hoisted myself out. The dirt from the grave sides felt dry and crumbly under my hands.

The light was in my face again. "Step aside."

Planetary Residential Commission agents had no respect for the dead. I was tempted to teach them some. "Leave. Now."

PRC hands grabbed at me, holding me, patting me down. One of them took my knife. The other grabbed my breast and squeezed.

"No tits," he said as he slid his hand down between my thighs. "But feels like she's got a nice, tight slash."

Mistake number one.

No female plays pro without getting groped in the locker room now and then. I'd been hit on for eight straight seasons, usually by rookies or new trades who hadn't been warned. If they came back from injured reserve, they never touched me again.

Nobody put their hands on me.

I took out two of them with one leg sweep, forcing the first backward and down by the hair while cracking some of the second's ribs with my boot. The third came running at me from behind, and I turned so he could collide with my fist. His nose fractured under my knuckles, and his jaw would have been next, but the fourth dove between us and tried to knock me away with a shoulder to the center of my chest.

It wasn't much of a block—I'd plowed through much worse on the field—and I weighed too much for him to move me. The problem was the not-at-all-human thing that swelled inside me. Worse than anger. My vision sharpened, my mouth dried, and I felt my claws punch through my gloves. Something huge and vicious lived inside me, and now it wanted more than a fistfight.

Take them down.

Rip their bellies open.

Fill your hands with their guts.

Take them down, Jory, now now NOW.

I fought it, curling my hands over, cutting my own palms as I forced my claws back into my fingers. Denying the ferocious surge was like being scalded from the inside out, but I'd kill them all if I didn't. Rij had taught me breathing methods to get through these rage spells, but that was for the game. With my mother there, it was personal.

And suddenly I wasn't too sure I could hold back the beast.

I had to get away from them. Right now. Before I could pick a direction, the fourth agent fumbled in his jacket, produced a weapon, and fired.

Mistake number two.

Light and pain crackled over me. From the whine of the blast I knew his weapon had been adjusted to heavily stun almost any life-form. Any life-form not wearing insulating thermal wraps, that is. In my case, it was like getting stung by a great big bee—it just pissed me off more.

"That didn't work, did it, asshole?" I knocked the pistol from his hand, grabbed his jacket, and jerked him up off his feet. As he dangled, eyes bulging, I showed him all of my pretty teeth. "Want to try again?"

I might have done more to Mr. Trigger-Happy, but the fifth agent stepped up to me. He was tall enough that he didn't have to stretch much when he put his gun to my head.

"Mine is set to burn a hole through your brain," Bright Boy told me. "Your move."

I dropped the terrified man and enjoyed the subsequent thud and grunt. "Tell them to keep their fucking hands off me."

"All right." He held out a hand. "I want to see some ID."

I tossed it at him as his men picked themselves up and brushed off their tunics. Trigger-Happy muttered a few nasty things, but Bright Boy told him to shut up and everyone to back off.

He checked my ID, gave it to one of the others, went over to the grave, then came back to me. "Identify the cause of death."

I was briefly tempted to say something like bubonic plague, but then they'd probably make a huge deal out of it and quarantine the entire commune. Besides, they'd caught me; it didn't matter that they knew.

So I told him the truth.

"Holy shit," one of them muttered. He had his light trained

on my ID tag. "You know who she is? Jory Rask. The runback with the NuYork StarDrivers."

"I don't care if she's the first lady." The one in charge ran his beam up to my face again. "Remove the eyewear and show me your hands."

My sunglasses and gloves were all that stood between me and being exposed for what I was. When I was a kid, Mom had even sewn my shades on a strap around my head, to make sure they didn't come off while I was playing.

No one must ever see your hands or your eyes. Promise me you'll be careful, Jory.

I took them off.

Mistake number three.

Someone spit on the ground. Trigger-Happy muttered more filthy words.

"You're under arrest." Bright Boy turned to his men and pointed to the grave. "Drag that thing out of there."

The PRC took me back to the city. Being secured in the back of one of their unmarked vehicles proved convenient—no drone monitor inside—so I took the opportunity to conceal a few items. Once we arrived at their regional office, they did the obligatory body-cavity search (and missed everything I'd hidden), gene print, uterine check to see if I'd given birth, then instantly revoked my planetary residential status.

I gave them as much trouble as I could without actually killing anyone.

After trashing three drones and spoiling a few more PRC profiles, I ended up cuffed to a chair in an inquiry room. The interrogator who finally came in was a blocky, middle-aged man with a blunt face and oddly small, delicate hands.

"Jory, Jory, Jory." He sounded like a disappointed parent. "Now we know the real reason why you're the fastest runback in the game."

"You figure that out all by yourself?" I asked him. "Or did you need a drone to break it down on a spreadsheet for you?"

"Alien blood always shows through." He sat behind the console in front of me. I returned his pleasant smile with an unblinking stare until Prissy Hands tossed my shades in my lap. Someone had been thoughtful enough to smash the one-way lenses. "Did you really think you could hide behind them forever?"

Besides being nearly seven feet tall, I had six fingers on each hand, and my mother's thick, matte-black hair. Unlike hers, mine curled, so I kept it short. All of that still passed as Terran.

Except my eyes.

I shifted, let the ruined shades drop to the floor. "Worked for twenty-four years."

"The gloves were clever. From the way they were fitted, no one could tell you had an extra finger on each hand." He took out a cigarette and lit it, probably to show me he wasn't above breaking the law. I'd heard they did things like that—establishing rapport with the filthy, disgusting half-humans before they tossed them off-planet. "Did she have your epidermis bleached?"

Couldn't he tell that I was naturally Terran-skinned? "No, genius, she used beige spray paint."

"You made the early-morning news." The agent blew some smoke in my face, then tapped a panel switch.

A wall screen flickered on, showing the image of a seething crowd around a bonfire. They were yelling and laughing and throwing stuff on the flames. Stuff like StarDrivers jerseys with my number, and photoscans of me.

I guessed that couple from Toronto would be getting their kid's name changed real soon.

"Not every day a World Game MVP turns out to be an alien crossbreed." He tapped some ashes into his cupped palm. "People are a little upset."

"People survive."

"You may have a little trouble. You've been fired from the StarDrivers for violating junta regulations. All your assets and property have been confiscated, and don't count on getting them back." Prissy Hands switched off the vid. "I don't have to tell you the rest. That alien mother of yours probably made you memorize the law."

Sure, I knew the law—nonhuman life-forms were denied residence status on Terra. Especially any progeny—generally referred to as crossbreeds—resulting from human/nonhuman liaisons. Terrans didn't want us polluting their precious gene pool.

My existence had certainly muddied the waters in a big way.

"I know this has been a tough night for you, but we're not really monsters, you know. I'm just pushing paper and doing my job." He gave me another let's-be-pals smile that didn't reach his eyes. "I'll see to it that you get some compensation, enough to help you make a fresh start off-planet."

"And you want what for this incredible act of generosity?"

"Information." He leaned back in his chair. "Tell me about the underground, Jory."

So this was how they did it. They caught one crossbreed and bought their way with credits to the rest. Made sense.

They weren't going to buy me, though. "It's deep, dark, and has a lot of rocks in it."

"We know you purchased the house two years ago. No legal residence on file prior to that." He slowly exhaled smoke, making it float in little rings between us. "Someone helped you, Jory. Someone kept you hidden somewhere all that time. Give me a name and I can make good things happen for you."

"Okay." I sat back, too. "MacDonald."

"MacDonald who?"

"Old guy." I yawned. "Liked hanging out with a lot of nonhuman types, you know."

Prissy Hands got busy and took notes. "Where does he reside?"

I pretended to think. "On a farm."

"Where is it located?"

"I can't remember . . . oh, yeah. Eeyigh-eeyigh-yo."

He frowned at me. "What is that? Some Indian reservation?"

Christ, he didn't get it. He was *that* stupid. "Let me spell it out for you again." I leaned forward. "Old MacDonald. Had a farm. E-I-E-I-O."

He looked as if he might lunge across the desk for a moment, then he shoved the notepad aside. "I want the nest, Jory."

The nest. Like we were rodents. "Go climb a flagpole. Maybe you'll spot one from up there."

He didn't like that either, judging by the color he turned. "You fem breeds are all the same, think nobody can faze you. Then you get to transport and start wailing and begging to stay."

Compared to junta training camp, deportation was a cakewalk. "Yeah, well. Don't hold your breath, you jerk."

"We'll see." He waited a minute. "Last chance, take it or leave it."

He bored me, so I started running play patterns in my head.

Prissy Hands got out from behind the console. "Jory Rask, you've been charged with two violations of the Genetic Exclusivity Act." He dropped the cigarette butt and ground it out under his footgear. "Your deportation arraignment will be in one hour."

That fast? "I want counsel."

"You're a breed." He took off his jacket and rolled up his sleeves, revealing muscular forearms. His girly hands really looked ridiculous attached to them. His smile finally crinkled the skin around his eyes as he loomed over me. "You don't get one."

Then he started hitting me with his small, hard fists.

Exactly sixty minutes later, I stood before a drone judge. One of my eyes had swollen shut, but I could still see out of the other one. Not that there was much to look at. Because of the media blitz on me, the drone judge had barred the public and the press from the courtroom. Everyone else kept their distance.

Don't want me getting my illegal blood on them.

They'd scrubbed my face before marching me in, but the cut on my cheek kept oozing, and my belly hurt so much I couldn't stand up straight. The agent hadn't broken anything, though—if he had, I'd have qualified for a medical postponement, and he didn't want that.

Have to get the disgusting alien crossbreed off the planet as fast as possible.

"Docket number two-seven-one-four, *Terra* versus *Jory Rask*," the bailiff read as the prosecutor took position beside me. "The charges are aiding and abetting an illegal alien, and maintaining unlawful planetary residence status."

I needed to spit out some blood. *Where the hell do you spit in a courtroom?*

The prosecutor, a petite blond female with an immaculate suit, gave me a single sideways glance before inserting a legal chip in the console. "The State asks for standard sentencing, Your Honor."

The drone judge processed that, then asked me, "Defendant Rask, are you aware PRC statute number six-eight-four automatically suspends your right to representation and trial by jury?"

Blood and saliva dribbled over my split lip as I muttered, "Yeah." I spotted a small wastebasket at the end of the console, and made use of it. "I'm aware."

The prosecutor made a disgusted sound and put another two feet of console between us.

"Jory Rask, this court hereby sentences you to immediate deportation from the planet Terra." The recorded sound of a gavel striking wood echoed through the nearly empty courtroom. "Next case."

A middle-aged, overweight woman switched off the drone recorder in front of the bench and removed the case file chip.

That's it? I'd expected it to be quick, but not fifteen seconds.

The prosecutor immediately left the courtroom, doubtless to visit a lavatory and spray herself down with disinfectant.

I didn't move. I had one last thing to do, and I wasn't leaving until I did it. Even if that meant another beating. "I have a request."

"Sentence has been passed." The drone's fake face turned to the bailiff. "Remove the convict."

"I want my mother's body," I said, louder.

"Shut your mouth, breed." The bailiff came to haul me away from the monitors.

I hiked up my good leg and kicked him away. The fat woman holding the case file chip screamed. Over the noise, I repeated, "I want. My mother's. Body."

The drone judge clicked and whirred as it buzzed over the big desk and hovered in front of my face.

"You are found to be in contempt of court," it told me, displaying whatever penal code I'd violated. "A fine of one thousand credits will be levied against your estate prior to deportation."

"Where is she?" I breathed through the pain kicking the bailiff had sent up through my battered torso, and wrenched my arms apart. The chain between my handcuffs snapped, freeing my hands. "What did you do with her?"

"We put the dead alien in a disposal unit."

I turned around and saw Pretty Hands standing behind me. He was smiling again. Like he had the whole time he'd beaten me.

The bailiff had gotten up and grabbed me, but I shook him off. My claws emerged, and the fat lady shrieked even louder as she ran from the courtroom. I waited until the screamer exited before I asked, "Why?"

"It's what we do with trash. Here." He threw something old and silver at me. "A memento."

I caught the necklace of tarnished links without thinking. Mom's vocollar. The one she'd never removed, even though it wouldn't translate anything away from a Jorenian linguistic database. I'd left it on her neck, intending to bury it with her.

"Rask." He drew his weapon and trained it on my heart. There were new bruises on his knuckles now, along with a deep bite mark. He got the latter when I demonstrated what I'd do if he tried to mouth-rape me. "Want to join her?"

For a minute, I was tempted. What did I have to live for? Everything I loved was ruined, dead, gone. My mother. My home. My career. My reputation. The only thing I truly cared about had ended up in a waste processor.

This is what it must have been like for her. From the moment I was born. How did she handle twenty-five years of exile from everything she loved?

The pendant suspended from the heavy links cut into my fist, and I uncurled my hand to look at it. The silver and black Jorenian pictograph was the symbol for the number seven.

I'd forgotten about them, too.

I looked at the PRC agent, who still had me in his sights. Somehow he seemed smaller, more pathetic.

"No, thanks." I slipped my mother's vocollar over my head, felt the links settle, cool and weighty, against my skin. "I've got someplace to go."

CHAPTER TWO

"An unaltered path is both boon and affliction."

—Tarek Varena, ClanJoren

Being deported would have been a swift, no-nonsense business, had I not been a celebrity athlete. The PRC agents assigned to haul my ass to Main Transport tried to sneak me out through the judges' private entrance, but the media sharks were already teeming and hungry.

As we emerged outside the courthouse, a wall of photoscanner lights popped, blinding me, and a thousand voices called out my name and the inevitable, brainless questions.

"Jory! Over here!"

"How did you fool the junta for eight seasons?"

"Did your coach know?"

"Jory! What do you have to say to your fans?"

The lights made it hard to see, and I stumbled a few times. The PRC agents ended up half carrying, half pushing me through the media and down a narrow corridor formed by security barriers and police bodies.

The spots faded from my one good eye, and I saw my fans were also waiting—hundreds of them, packed like sardines on both sides of the barriers, dressed in their StarDrivers tunics and jackets. Some of them carried handmade signs. As soon as they saw me, they began yelling, just as loud and raucous as they had when I'd brought in the final sphere-down in World Game XXXIX.

None of them wanted my autograph now, though.

"Alien scum! How dare you!"

"Disgraced the game!"

"Get off our world!"

"Filthy breed!"

When I passed a man pressed against two cops and the barrier, his arm shot out and landed an off-center punch to the side of my head. I lunged, but the agents dragged me back.

"You bitch!" the lousy puncher screamed. "You fucking alien bitch! How could you do this to me?"

I looked, saw who it was. "Oh, bite me, Coach."

As we walked through the gauntlet, the fans began spitting on us. The PRC agents swore and shielded their faces with their hands, but I kept my head up and didn't allow myself to flinch. By the time we reached the glidecar, I was literally dripping with saliva and mucus.

"Here." After shoving me in the back, one of the agents tried to hand me a handkerchief.

"What are you, going soft?" Another one snatched it away. "Let her remember what we think of her kind."

My skin crawled with disgust and humiliation, but I only used my sleeve to clear my eye and wipe off my face. He was right; I needed to remember this.

Once security cleared the road, the PRC drove me directly to New Angeles transport. On the way, I was informed that all of my personal assets had been confiscated and were now the property of the PRC. They allowed me only enough credits to buy passage to the next inhabited solar system.

"And you shouldn't come back," the agent told me. "Because if you do, this will seem like a bon voyage party."

I wasn't the only one being deported that day, either, as I found out when they loaded me onto an orbital shuttle.

"What'd they snatch you for?"

I looked down two rigs, and the other prisoner grinned back at me. Judging by the size and shape of his blunt tusks, he was a Nivid. Just my luck. I'd run into them a few times—they were rabid sports fanatics—and they never knew when to keep their mouths shut.

As he immediately demonstrated. "Beat the waste out of you, huh? *U'flargot* PRC, always get off on pounding us."

That seemed a little too warm and fuzzy, even for a Nivid. He sure as hell didn't have any bruises, and he didn't know me.

A grunt erupted from him as he shifted in his rig. "Ought to be a law protecting *us* from *them,* instead of the other way around."

On the other hand, maybe he did. I thought about it as I watched the viewport and the dwindling sight of my homeworld.

"Been down there long?"

I already knew that for the right info, the PRC would pay any alien or crossbreed a little good-bye bonus. *And who better to cozy up and leech me than a chatty sports fanatic who just happens to be getting kicked off Terra at the same time?*

When I didn't answer, the Nivid decided to tell me his life story, which was almost as tedious as he was. I ignored most of it, until he got to the part about trying to join the alien underground.

"—heard they were all over in New Angeles, but try making a contact, I mean, paranoid doesn't even come close—"

If he was a spy, he stank at it. If he wasn't, he was an idiot. Either way I was tired of his mouth. "Shut up, moron."

"Only trying to make conversation." He attempted to stretch out one of his stubby legs, but the restraint harness kept him pinned. "Don't know why they have to keep us rigged like this. Not like we're going to jump shuttle, right?"

I certainly would have tried, just to shove him out an air lock.

"So what's your story, lady? Get in a fight with a horny Rilken?" He huffed out oinking laughter.

My restraints were loose enough to let me lean over a little. "I gutted someone who looked like you."

He stopped laughing. "Why?"

I gave him a slow, nasty smile. "I don't like tusks."

That kept him quiet until we docked with the passenger shuttle.

"If you attempt to return to Terra, you will be immediately detained and prosecuted," one of the guards told us just before we were pushed through the connecting portal. "The sentence for repeated violation of the GEA is twenty years, dome labor."

There was nothing on Terra worth two decades of excavating basalt so rich people could move into exclusive vacation habitats on the moon. "Don't leave the light on for me."

The Terran steward on the shuttle didn't remove our restraints until we had left planetary orbit. "Captain wants to see both of you."

The Nivid backed away as soon as my hands were free. "If it's not too much trouble, can she go first?" The steward shrugged. "Thanks. Where's the bar? I need a drink."

The captain received me in his office, a spartan compartment off the helm. I was in luck; he wasn't a full-blooded Terran. Someone in the family, probably one of his grandparents, had been an avatar.

He walked a circle around me. "You look like walking waste."

I didn't smile; my face hurt too much. "No time for makeup and a shower this morning. You know how it is."

"The usual spit-bath before mandatory deportation, eh? Well, I may be a breed myself, but don't let the feathers fool you," he said as he folded his stunted wings and settled behind his desk. "I'm not operating a charity route."

That sent my first appeal right out the viewport. "Right."

"I run regular jaunts to Terra, so I won't employ you, either. Your kind jump ship, and I'm not getting hit with conspiracy charges." He tapped a few keys on his console. "Standard drop is on Andromeda in Alpha C; that's as far as what's left of your credits stretches." He hesitated. "The colony there's pretty rough, but if you're smart, you can earn your way out in a few cycles."

Peddling my ass at Main Transport, no doubt. "I'll pass." I took the ring I'd hidden in the knotted mass of my hair and placed it on his desk. "How far will this take me?"

"Son of a fem." He picked it up, rubbed off some dried spit, and examined it. "This is a World Game ring." He scanned it. "Sea starlite, yellow diamonds, platinum setting. Not bad. Not worth a fortune, but—"

"It's a Game ring." *My* Game ring, and I'd spent two weeks flat on my back recovering after that particular bloodbath. "You can get at least thirty thousand for it, forty from a collector. So how far?"

He checked the inscription, and hooted with pleasure when he saw the MVP. "Where do you want to go?"

I had to get to the others. "Joren."

"Not even for ten of these." He shook his head. "Too far off my route, and they're too damn twitchy about Terran traders."

I wasn't a trader. "The nearest Jorenian vessel."

"Maybe, if I can find one." He thought that over, then squinted at me. "How much do you know about them?"

"They're tall, blue, and apparently"—I held up one of my six-fingered hands—"counting to twelve isn't a problem."

The captain uttered a single, chirping laugh; then his expression turned serious. "You should know there's bad blood between them and the League. A while back the Jorenians gave sanctuary to some clone who escaped from a Terran lab. Heard they adopted her or something. Anyway, when the League tried to take her back, the Jorenians got angry and broke treaty over the whole mess."

I folded my arms. "I'm not interested in making a new treaty with them."

"You don't understand. They broke with the *entire* League—every single member world. Pulled all their pilots and navigators off every ship, colony and science station, even the ones only remotely allied to the League. Since then they've had nothing to do with any League world, particularly Terra." He made a small gesture. "We've been told to give them a wide berth."

Avoiding Jorenian ships wouldn't get me very far. "But you will let me transfer to one of their vessels when you locate one."

He didn't look happy. "If they'll take you."

"That's all I ask." The Jorenians couldn't be as bad as all that, not from what Mom had told me about them. And I was one of them—sort of. "Thanks, Captain."

Before I got to the door panel, he said, "One more thing."

I turned. "What?"

He held up a flashy StarDrivers pennant. "Sign this for me?"

After my interview with the captain, I was shown to a cramped, dismal-looking cabin, where I spent an hour in the cleansing unit, then another icing down my knee, sterilizing my clothes, and tending to my wounds.

I'd made a good bargain on signing the pennant, I thought, eyeing the knife he'd given me in trade. The captain had offered a pulse pistol first, but I requested the blade.

"Pistols need recharging," I'd told him as I'd examined the Omorr challenge dagger he'd produced. Seventeen centimeters long and six millimeters thick, it had a clean, simple sweep from edge to tip. The high-carbon alloy was also better than anything I'd ever used on Terra, but I didn't tell him that. "This only needs sharpening."

"No knife fights on my ship," he warned as I left.

You don't fight with a knife, Jory, Rijor had told me when he'd given me my first blade. *You kill with it.*

"Right."

I put the knife aside, fixed myself a meal I could only pick at, then curled up on the too-short sleeping platform. I kept hearing my mother's voice, on my hotel console. She always sent her signals audio-only, but the last one had been too short.

Come home, Jory. I have not been feeling well.

I closed my eyes, reaching for the calm Rij had always told me to find whenever I grew angry. *Why didn't she tell me how sick she was?*

Calm didn't want anything to do with me, and the sounds of movement outside in the corridor kept distracting me. After an hour, I swore, kicked the linens onto the deck, then got dressed.

In the corridor I bumped into a couple of Tingalean snake-people, apologized, and asked if they knew where the bar was. They bared some fang and slithered past me without replying. I checked my wristcom, and it seemed to be working.

So what's their problem?

Two more alien passengers crossed my path, and likewise ignored me. I ended up accessing a public console to view the ship's schematic.

I sniffed at my tunic, which smelled clean. *Must be my pretty face.*

At the bar I spotted the Nivid, a little drunk and happily chatting up one of the big-chassised bar drones. Which reminded me, I needed to tell him to keep his trap shut about the underground. As soon as he saw me headed his way, however, he grabbed his server and jumped out of his seat.

"Here, it's yours."

"Relax. I don't want your table."

"Sure, sure." He sat back down and covered his tusks with one split-digited hand. "Anything you say."

Shadows fell between us, and I turned to see a couple of soldiers in brown uniforms approaching. One was a tall, pale-skinned humanoid with four extra arms; the other looked like an overgrown purple shrub. Both carried pulse rifles and nasty expressions. They also were weaving, enough to tell me they'd been at the bar about as long as the Nivid.

"Terran."

Everyone within ten feet of us simultaneously got up and took their drinks to the other side of the room.

I turned my head, gave them the once-over, and turned back to Tusk Face.

"Hey, I'm talking to you." The bushy one prodded me in the back with something hard and leafy. "Get out of here."

His buddy's cheeks hollowed as he hooted something my wristcom translated as "Now."

I thought of the blade, which I'd clipped in its sheath to my belt. What would I tell the captain? *It isn't a knife fight if the other guy doesn't have a blade, sir.* "Why should I?"

"We don't want you spitting in the drinks." The shrub extended a couple of arms—or branches, I wasn't sure which—and gave me a good shove. "Go on, get out."

I shoved back. "I'm not your goddamn waitress."

The hooter pulled his rifle off his shoulder and tucked the

back end under one of his arms. Whatever he said this time was either so obscure or so obscene it didn't translate at all.

The Nivid jumped up. "I should probably go."

I clamped a hand on his shoulder and pushed him back down. "These morons friends of yours?"

"Not at all." He stared past me at someone else. "I don't want any trouble here."

"I'm glad to know it," a pleasant voice said. "Sergeant, unless you want to spend the war hacking out latrines on a detention moon, you'd better shoulder that weapon right now."

I glanced at the third soldier who'd joined our happy little group. This one wore an immaculate brown-and-red uniform, rows of commendation ribbons, and a major's insignia gleaming on his stiff collar. He also had a brown pelt, warm, intelligent brown eyes, huge ears, and two slitted nostrils in the end of his powerful-looking muzzle.

The eyes said *puppy* but the teeth said *wolf*.

The major studied me for a moment. "As I live and pant. You're Jory Rask, aren't you?"

Before I could answer, the Nivid knocked over his flask, spilling spicewine all over the table. "You're her? Jory Rask? The NuYork StarDrivers runback?"

Even the two soldiers jittered back a step or two, looking stunned.

"Yeah, that's me." I righted the Nivid's flask. "What do you want? An autograph?"

"I'd rather buy you a drink." The major turned to his men. "You two have something better to do. Go and locate it."

The shrub and the hooter staggered back to the bar. The Nivid was still sitting in the only other chair, staring at me and muttering, "Jory Rask?" under his breath.

I leaned forward so only he heard me. "Go play with a junker, Tusk Face. And don't run your mouth about the underground anymore, or I'll sew your lips together."

Terrified, the Nivid bobbed his head, grabbed his drink, and fled.

"What's his problem?" the major asked me as he took the vacated seat.

"Got me."

He gestured for the nearest drone, who sped over to mop up the mess. "Carafe of spicewine okay?"

Synalcohol had little effect on me, thanks to my bad blood, so it didn't matter what he plied me with. "Whatever."

"Don't say that." He slapped a paw over his chest with a dramatic flourish. "I'd love to have your offspring. How about we get intoxicated, go back to my place, and you impregnate me?"

Aliens. My mouth hitched. *No wonder they make Terrans so nervous.* "Doesn't work that way with my kind, pal."

"Pity." Amusement glittered in his eyes. "Major Thgill, Allied League Border Patrol, Engineering Division."

"Jory Rask, homeless deportee." I waited as he ordered the carafe from the drone, then asked, "Kind of far from Pmoc Quadrant, aren't you?"

"I've been on leave to see my parents. They've just retired to Europa Station." He nodded toward the soldiers huddled at the bar. "I'm sorry about that. My men usually know better, at least when they're sober. I'll see it they don't harass you in the future."

"I'm not worried." I took a server from the drone, then watched as it served him his. "They really think I'd spit in the drinks?"

"Terrans tend to ejaculate a lot of saliva around nonhumans." Thgill gave me a wolfish grin, and added deliberately, "Present company excepted, I hope."

Terrans loved to spit on aliens, as I'd discovered leaving the courthouse. And now the aliens were ready to kick my ass, thinking I'd do it to them. Yep, jaunting to Joren was going to be a real treat.

Thgill misinterpreted my silence. "I'm just kidding, Jory; don't let it bother you. Not everyone hates humans at first sight."

"How many have you hung out with?"

He made a face. "Well, okay, your kind are . . . fairly disagreeable. Maybe you could change that, you know, provide a more positive example."

I pulled off my shades, then my gloves. "I'm not exactly a role model."

Thgill whistled. "Very nice." He peeked over at the front of my tunic. "Hiding any fur under that, runback?"

"No, sorry."

He shrugged. "Can't blame a male for asking. So tell me, what was it like, playing N.A. for World Game?"

Thgill kept me company over the next couple of days. He was amusing, harmless, and played a decent game of whumpball. He also talked a lot about other species and quadrants, and how they figured into the Hsktskt conflict.

"See, if we could recruit a few battalions of Tingalean warriors, we'd have it made." He turned from our table in the galley and nodded toward the pair who remained aloof from the rest of the passengers. "They're fast, fanged, and even their blood is pure poison. The lizards wouldn't have a chance."

I took a bite of my synpro sandwich and studied the two snake people, feeling very glad I hadn't gotten nasty before. "Why don't you sign them up?" I asked after I was done chewing.

"They're pacifists." He made a rude noise to illustrate his opinion of that. "Don't believe in open aggression, especially against other reptilian species. Just like the Jorenians."

I drained my coffee. "How much do you know about the Jorenians?"

"They were some of best pilots and navigators we had, until they broke with the League. Want more of that brew?" At my nod, he got up and refilled our servers. "Damned pity we lost them, too," he said when he sat back down. "Get the Jorenians riled, and those big blue bastards make the Tingaleans look like a bunch of garden pests." He eyed my coffee. "How can you drink that stuff black?"

"Because if it were green, it would make me throw up. So

they're siding with the Hsktskt?" If they were, I'd just skip my visit to Mom's homeworld.

"They don't side with anyone that I know of. Completely neutral—militantly neutral, in fact. They've declared their home system off-limits—not that any League ship would go within a hundred light-years of them after what happened with the fleet in Varallan." He thought for a moment. "I heard they joined up with the Aksellans to free some Hsktskt slaves on a depot world, but it sounded like a one-shot deal." He gave me an uneasy look. "You know, Jory, they say a Terran made them break treaty with the League."

"Yeah, the captain told me about that squabble over that escaped clone. So?"

"They don't trust Terrans. You *look* Terran."

"Should make my visit interesting." My knee felt like it was swelling, and I got up slowly, handing my tray to the nearest server drone. "I think I'll hit the platform early. Night, 'Gill."

Ever the gentleman, Thgill insisted on walking me to my quarters. "You're limping again," he said along the way.

My whole leg had been aching all day, but now it pounded. "Old game injury."

He tried to take my arm. "Have you seen the ship's doc?"

"I'm not a cripple." I tugged away. "Forget about it."

Thgill wouldn't be sidelined. "Listen, if you won't report to medical, at least let me take a look at it. I've had some first-aid training."

We reached my door panel. "You just want to see the inside of my cabin," I said, and keyed the door to open.

"Am I so obvious?" He caught my arm as I stumbled over the threshold. "*Suns*, Jory, you're in pain. Come on, you know I'm not going to jump on you. Not without an invitation, anyway."

I sighed. "All right, get in here."

I stripped off behind a privacy screen and shrugged into a robe, then retrieved the tool kit I'd bought off a crewman. Amazing what people will trade for a couple of signed pennants.

Thgill came over to watch me sit down and unwrap my thermals. "Holy High Bitch."

I raised my brows. "Now *that's* one I haven't been called before."

"Not you, sweetheart, one of my deities." He knelt in front of me, reaching to touch, then pulling his hand back. "Why did they do this to you?"

"To keep my leg attached, mostly." I pulled out a calibrator and went to work adjusting the brackets. The surrounding tissue was badly inflamed—again—so I'd have to start another round of antibiotics. "Grab some ice out of the prep unit for me, will you, 'Gill?"

My knee was ugly—a tangle of scarred flesh and tech—but at least it still functioned. The original had gotten pulverized five years ago, during my rookie season. The underground fixed me up with a biomechanical replacement, but shockball wasn't a gentle sport, and I'd gone downside for ten more retrofits since then. After the last one, the underground's doc had lowered the boom: one more bone shave and I'd lose my leg from the knee down.

"The Terrans are finally starting to catch up with you, Jory. Same ones you ran circles around last season," he'd said when I'd argued with him. "Go out with a little dignity, will you?"

That had been a month ago, when I'd started my retirement plans to take Mom and move up into the mountains, where no one would ever find us. Now I had dignity. No mother, no money, no place to live, no prospects outside of keeping a promise, and maybe reduced to begging help from six complete strangers who owed me nothing, but all kinds of dignity.

"Here." Thgill had fashioned a pack out of the ice and a piece of cloth, and applied it to my leg. "How long have you had this prosthesis?"

"Eight years." I took out the syrinpress and dialed up what I needed for the infection.

He lifted the pack for a moment to examine it again. "I've never seen components like these used on a living being before. Kind of scary to think someone would."

"Recycled cybertech. Some of my friends in the underground steal it from service drones. I take amolynicillin for the occasional infections," I added as I pressed the port against my neck and infused myself. The sting barely registered. "Don't look at me like that. I couldn't exactly check into a hospital."

"I'm just trying to imagine someone like you in hiding."

I told him a little about it, although I skipped over the years we'd spent underground. I liked 'Gill, but he belonged to the League, which had gotten into the war with the Hsktskt because some Terran had spoken out against the slave trade. Instead I told him about the old-fashioned future-agers commune Mom and I had taken sanctuary with.

"They liked living off nature, burning herbs, and remembering past lives. We didn't belong to the cult, but since farm equipment and hypnotherapists are expensive, I gave them money, and they tolerated us." I bent my leg, winced, and straightened it again. "Bribing them bought us a decent amount of living space, supplies delivery, and even a little protection for Mom when I was on the road."

He shook his head. "I can't believe the junta never caught on to you."

"Most shockjocs have their own private cutter, so I didn't have to use the team doc. People saw what I wanted them to see." I tapped my shades. "I had to get all my meds and tech replacements through the underground, and sub my blood and urine samples whenever there was a drug sweep—but otherwise, everyone left me alone."

"So how did they catch you?"

"Mom got sick, and died before I could get home. Someone at the commune decided the tip-off comp was too good to pass up and called in the PRC."

"What did she die of?"

"Chicken pox."

I remembered how I'd rushed home, only to find Mom's body on the floor of her bedroom. The brief, insane hope it was some kind of tasteless future-ager practical joke. Turning her over, seeing the horror that had been her face. Realizing my mother had died of a childhood disease that could have been cured with a single dose of over-the-counter inoculant. I'd carried her to the bed, laid her out.

Then I'd kicked a hole in the wall, which was what had screwed up my knee.

"She must have contracted it from one of the kids in the commune," I told him. "She'd let them in the house sometimes when she was lonely. Fed them tea and morning bread. I told her not to, but she never listened to me."

"At least she didn't suffer."

"I ran the house security vids." I didn't look at him, but my voice went flat. "She spent the last day convulsing from the fever and screaming my namc."

"Sorry."

"Yeah." I stared at my knee. "Me, too."

Thgill took out a scanner and made a couple of passes over my leg. "Jory, I design and build combat drones for the League. I don't know how, but you've already developed a tolerance to tech that's supposed to be incompatible with living flesh."

"Thanks, that makes me feel better."

He patted the side of my leg. "No, what I'm trying to say is, if you're willing to let me tinker on you, I think I can do better than this."

I was not going to hope. "How much better?"

"With the gear I've got with me?" He sat back and thought. "I don't have the interior femoral and tibial components—you need an ortho surgeon to put those in, anyway—but I can replace these obsolete supports, and recalibrate the ligament tension, range, and movement tolerances. Take a couple of hours in medical, tops."

What mobility I had was pretty precious to me. "Will I be able to run?"

"Your tech won't stand up to any more serious impact injuries, but as long as you don't play contact sports again"—he flashed his pointed teeth—"you'll run like the holy bitching wind."

I lifted the ice pack and hobbled over to the viewport. All around the ship, cold, black space stretched out for millions of star-pocked light-years. Terra was long gone. Just like shockball. And Mom.

All I have left is Joren. "When can you do it?"

CHAPTER THREE

"The path stretches beyond the House, beyond the world, beyond the very stars."

—Tarek Varena, ClanJoren

A few weeks later, I saw Thgill off at Kevarzangia Two, where the League had set up their quadrant command post. He wouldn't take any comp for the retrofit, but he made me swear to get in touch if the new tech failed.

"This is my personal relay code." He pressed a chip into my hand. "Memorize it, and you can reach me anywhere I am if you need help." He frowned down at my leg. "I still think you should go on-planet with me. K-2 has a decent FreeClinic; one of the docs there can finish what I started."

"No, I'll get it done on Joren. Thanks, 'Gill." I shook his hand, then, on impulse, gave him a one-armed hug. "Watch your back, okay?"

"Always." He gave my chin a friendly nip with his teeth. "Take care of my tech, sweetheart."

Three systems later, we still hadn't encountered any Jorenian ships. Naturally, my credits and the captain's generosity began to run out, and I was informed I'd better decide where I was getting off before the ship looped back for Terra.

I took a chance and got off at a Rilken outpost, and turned the last of my valuables into credits. I didn't have to look far for transport, thanks to a Rilken steward who saw me pawning my belongings. After a hard bargain with the nosy little demon, which bought me passage on one of the system trade ships, the *Chraeser*, which jaunted a munitions supply route to the front, and routinely encountered Jorenian ships on the way.

"Can't say they'll let you board if we do," the ship's steward told me. "Jorenians aren't real fond of Terrans."

"Yeah, so I've heard." I hitched my pack onto my shoulder and walked up the docking ramp. "Lucky me."

Life on board a gun runner proved a lot different. For one thing, its amenities made the Terran trader look like a pleasure cruiser. Basically what the *Chraeser* offered its passengers were cramped quarters, lukewarm meals, and zero recreation.

Because of their diminutive size, the Rilkens employed bigger life-forms to run their trade vessels. The crew were a hodgepodge of different alien species, and seemed interested only in staying on schedule, keeping costs to a minimum, and dodging battle space. If they said anything to me and the other passengers, it was usually a variation of "Get out of the way."

It did have one plus over the trader—a training deck. Since most of the other passengers were League soldiers or officials returning to the front to resume duties, the Rilkens had provided ample workout space and gym equipment on one of the cargo levels.

I started working out every day to keep my mind occupied and get my refitted knee in shape. Once we'd convinced the ship's doc to let us use the medical facility, 'Gill had done exactly as he'd promised. Within a few days of the retro, I had more range of motion and flexibility than ever before. The only problem left was the interior tech, which he had inspected and declared absolute junk.

"You get to a colonial hospital and have them replaced as soon as possible, Jory." He showed me the scans of the stress damage he'd found during his tinkering. "If you don't, one more serious injury and you're going to lose half a leg."

Which was exactly what the underground doc had told me—so much for the second opinion. I flexed my leg and ignored the clenching fear in my midsection. "They'll fix me up on Joren."

He gave me a suspicious look, then sighed. "If they don't, signal me."

That morning I pushed the future surgery out of my head and concentrated on getting back into shape. I figured if I couldn't make a place for myself on Joren, I'd get a slot hauling cargo on one of the trade jaunts. When I paused to up the resistance on the extension machine, a black blur passed by me, headed for the private rooms.

Huh. The Shadow's early today.

I'd seen the thing the first time about a week out from the Rilken docks. Tall, vaguely humanoid, and completely silent, it acted as if it didn't want to draw much attention to itself. The problem was, everyone noticed it—mainly because of the strange black garment it wore that covered it from head to footgear. Even more bizarre, the garment seemed to shimmer as it moved, blurring the outlines of its body. I couldn't even tell if it was male, female, or other. The thin, funny-looking black stick it always carried around was about a half meter long—too flimsy to inflict any real harm, and it certainly didn't need to use it like a cane.

Tall, Dark, and Indefinable showed up on the training deck frequently, always worked out alone in one of the private rooms, and never spoke to anyone. Often I got the feeling it was watching me, but when I looked in its direction, it was already gone.

I wonder where it's traveling to, I thought, then became aware that someone else hovered behind me. I lowered the weights and turned.

A large, hairy crewman with a sullen expression on the face sprouting between his legs pointed at the machine. "I'm using this now. Get off."

I'd seen him before on the deck. He was a bully who liked throwing his weight around with the smaller crewmen. He didn't impress me, although I thought if someone shaved him and taught him to walk backward, he'd make a passable, if somewhat rearranged, Terran.

"When I'm done." I went back to my extensions.

"I said"—he grabbed the collar of my tunic—"get off."

I let him drag me back, then pivoted, yanking free of his grip. A moment later I pressed the point of my trusty Omorr blade into his belly/neck, enough to penetrate his flight suit and scratch his skin.

I leaned in. "And *I* said, when I'm *done*."

The crewman swallowed, then nodded and backed off, hands out. I sheathed my blade and went back to the machine; then the hair on the back of my neck prickled, and I stepped to one side. The hairy crewman slammed butt-head first into the machine, yelped, and dropped over the bench, holding himself between the legs as he bled all over it.

Everyone on the deck, including the one in the weird black clothes, stopped what they were doing to stare at me.

"You can have it now." I went on to the next machine.

One of the soldiers on the deck escorted the semiconscious crewman to medical, while another wandered over to watch me.

"Speedy turn, Terran," he said. "Want to enlist?"

I pulled seventy-five kilos in resistance on the vertical bar, released, then recalibrated to add another twenty-five. "No."

The soldier looked up as the deck panel opened and closed, and Black and Blurry left. "He's been watching you, you know."

So he was a male, and I wasn't imagining it. "Who is he?"

"A killer."

I pulled down with the extra twenty-five, feeling the answering burn in my muscles. "I won't ask him for a date, then." I released the bar. "What's the deal with his clothes?"

"They're made of dimsilk; some bug on N-jui spits it out. The webs they make distort light. His kind always wears it." He shrugged. "Great disguise, but not much on body armor."

"And the little stick? What does he do with that? Scratch his back?"

The soldier rolled his one eye. "Why not ask him yourself?"

I got my opportunity the next day when the Shadow appeared by my table in the galley. I wouldn't have noticed him at all if he hadn't tapped the end of his stick on the rim of my server just as I was lifting it to my mouth.

I looked up at the blurry black shape that might have been his head. Maybe I was sitting in his spot. Maybe he didn't like Terrans. Maybe that stick contained poison-tipped spikes he planned to bury in my skull.

And maybe I'll get to try out my new knife again today. "What?"

"You move well," he said, in a voice rendered flat and metallic by some hidden translator device.

Another fan. Just what I needed. "Thanks."

"You not belong here."

Or not a fan. "So?"

He made a funny swirl in the air with the stick. "Dance with me."

Of all the things I'd expected to hear the Shadow say, that wasn't on the list. His translator must have shorted or something. "I didn't get that; say again?"

He repeated the same bizarre request. A Rilken passing by overheard it, stopped, and grinned up at me. "He challenges you to a fight."

"Oh." I looked at the Shadow's maybe-head again. "No, thanks."

"Moves only," he assured me. "No weapons."

How comforting. "Yeah, well, I'm not interested in fighting you with anything."

"Everyone fights." The little stick tapped the table. "Or dies."

"That's a very interesting philosophy." Or a very pointed threat. "Would you mind leaving now? I'd like to finish my meal."

"Wait." The Rilken tossed a handful of credits in front of me. "I'll pay to see a practice match between you and him."

It hurt to ignore that much currency—I was almost broke—but I shook my head, got up from the table, and walked away.

The Shadow still came to the training deck every day, and shut himself up in one of the private rooms. I had no idea what he did in there. Sometimes he emerged and stood against one wall panel, watching me. Playing ball made me used to having an audience, so I ignored him. Most of the time I thought about Joren and what I'd tell the others.

Mom had told me about the other ClanChildren of Honor—what her people had called the other six Jorenian crossbreeds who, like me, were born to survivors of the attack on the *Moon-*

Wave. She'd made me promise to find them someday and tell them about what had happened twenty-five years ago, though not everything—she'd felt that would have been cruel. Under the circumstances, I agreed. Now I knew exactly how they'd feel if they learned the whole truth.

I was thinking about that—and them—on the day I nearly died.

That afternoon I started as always, working my muscles from the feet up. I spent extra time on my knee, on the same machine the hairy crewman had tried to pull me off of.

So I go and I tell them about what happened with the Moon-Wave—*Hi guys, guess what? All of our mothers took a research trip into space twenty-five years ago. They were attacked, captured, and then some really nasty stuff happened to them—How are they going to take the news? What if they aren't on Joren? How am I supposed to track down—*

One minute I was pulling up on the leg grips; the next something made of alloy groaned and a flurry of indistinct black appeared next to me.

"Terran!" Light flashed in front of my face. Something silver and sharp. The Shadow knocked me off the bench, and the blade came down, right for my abdomen.

I didn't think; I reacted. Before he stabbed me, I rolled under him, scissoring my legs to knock him off balance. A fraction of a second before I would have, he wasn't there anymore.

What the hell—

The deck shuddered and my ears rang as something huge and heavy smashed down beside me. Sharp alloy fragments pelted me, and I closed my eyes and threw up my hands to protect my face. A good move, too, because a sharp, slicing blow sheared across the backs of my arms.

Cut me, the bastard.

I didn't wait for his second move, but got my feet up under me and pushed myself to my knees. Blood streamed down my arms, but I focused on the Shadow, who stood only a foot away, in a fighting stance. Instead of the stick he held two short daggers, one in each hand. What was left of the weight unit I'd been working out on lay on the deck in a million pieces. It didn't make sense.

He wrecked the machine, then *stabbed me?*

I rose, keeping my shoulders down and protecting my torso with my bleeding arms. I needed a weapon, something I could knock the blades away with or use on his head.

He spun the twin daggers in his hands, then brought them together, side by side at the hilt. My eyes widened as the blades seemed to melt together and suddenly became one knife, which he tucked away beneath the blurry dimsilk. "You bleed."

"No shit." I felt the wall panel hit my back and groped sideways for the equipment storage unit. He started coming at me. The moment I felt alloy, I grabbed and pulled out a heavy suspension bar and swung. "Your turn."

An inch before I clubbed him, he reached and snatched the bar out of my hand. He did that so fast that my arm nearly came out of its socket. "I not hurt you."

"Yeah?" My inner beast rolled, and I felt my claws emerge. If I used them, his insides were going to end up all over the deck. If I didn't, mine might. "What do you call this? Foreplay?"

"I cut free." He pointed to the remains of the weight unit. "Recoil."

I glanced down. One of the thin alloy cables from the leg grips had my blood all over it. He stepped back, and my claws retracted. "What the hell happened?"

A crowd had gathered by then, and an engineer knelt down by the unit. "Looks like the supports buckled. Good thing he shoved you off it, Terran. It would have crushed you."

"Let me see." I got down, dripping blood all over the deck as I examined the machine. Where it had snapped, the twisted alloy had a scorched look to it. "Someone's been screwing around with this."

I looked to see what my hero had to say about that, but he was already gone.

The captain of the *Chraeser* stopped by to see me in medical, and made it clear he wasn't happy with the attempt on my life.

"Next time I put you off the ship." He pinched his central nostrils closed as he watched the ship's doc finish the suture work on my arms. "You comprehend that?"

"Sure." The stink of cauterized flesh didn't bother me. I usually smelled pretty crispy after a game. "Thanks for caring."

"Next port is Ichthora."

I'd heard about that mud ball already—a real refuse heap of a planet. "I'll stay in my quarters. What about the weight unit?"

He uncoiled a tendril and rubbed the side of his slimy skull. "What about it?"

"Somebody sabotaged the support joints by using a torch on them. Maybe that hairy idiot who gave himself a concussion trying to slam me."

The captain shook his head. "That crewman left the ship the day after you fractured his skull."

"Then the guy in the weird black clothes."

Now he laughed. "You're not worth his time."

I stayed in my quarters after that, except for meal intervals. Since no one had personal prep units, and I wasn't trusting anyone on the ship, I still had to get food from the galley. Nothing happened. A few days passed; then I gave up hiding to sit and eat in the galley rather than haul my meals back to my quarters.

Even with the meal breaks, I was sick of staring at four wall panels, with nothing to do but think about Mom and Joren. One morning I got on the console and initiated a database search for the six Jorenian names Mom had given me. After a momentary hesitation, I added a separate name to the search string.

Kieran. My father. At least if he's dead or in jail, I'll know.

Since the console was older than me, it was going to take some time to complete the search, so I headed out for refueling. In the galley I prepped a simple breakfast and took my tray to an unoccupied table. Thinking about Kieran—about finding Kieran—made my stomach clench.

My common sense gave my stomach a good mental kick. *Quit dodging it. You're out here; he could be out here. Better to find out where if you can.*

A familiar stick appeared on the edge of my table. "Terran."

"Not again." I glanced around the Shadow. The Rilken crewman and a couple of his buddies stood gathered in a loose ring behind him. "What now?"

"Dance with me."

Again with the dancing. I rubbed one temple. "Look, pal, I told you, I'm not going to fight you. Besides, the captain will dump my ass on a swamp world filled with face-suckers if there's any more trouble involving me and blood on the deck."

"Captain works for me," the blur said. "Does what I say."

I eyed the Rilken, who nodded. My arms had barely healed, but if I didn't soon make some money, I'd be landing on some other dismal mud ball anyway. "Okay—moves only. No blood, no weapons, no killing." He nodded. "When?"

He stepped back from the table and pointed toward the corridor with his stick.

The fight didn't start right away: The Rilken had to clear out the training deck, set up a quad, alert the crew, and take the bets. A lot of currency exchanged hands, and plenty of eyes followed me as I went to warm up.

"You've got testicles, fem," one of the crewmen said, giving me a friendly slap on the back, then tagging along with me to one corner of the quad. He was a big, skinny humanoid with four extruded cranial stalks. Eyes like bunches of Terran grapes clustered on the end of each stalk.

"Not when I checked in the cleansing unit this morning." I stood beside one corner of the quad and stripped off my outer tunic.

"You know what I mean." He nodded toward the assassin, who was standing on the other side of the quad. "Not many *males* would take on a blade dancer."

My tunic fell out of my hands. My jaw fell out of my face.

"You didn't know?" The crewman rotated his eye clusters toward the Shadow, then coughed. "Gods, you're in for it now."

The Rilken who had set me up crouched in one corner of the quad, going over bets recorded on a datapad. He became my target. His datapad went flying as I yanked him out of the quad and pinned him against the outer ropes, and the shocked look he gave the distance from his feet to the floor was comical.

Or would have been, if I hadn't felt like strangling him with his own tentacles.

"See that guy over there?" I forced the Rilken to look at my opponent, then at me. "You didn't mention that he's a blade dancer."

"I thought you knew," he said, panic and my grip making his voice squeak.

"Bullshit. You thought I was stupid." I wrapped a hand around his scrawny throat. "No one in their right mind would spar with a dancer."

"Terrans are crazy; everybody knows that," he said, wheezing a little when my hand tightened. "Sorry, sorry. He asked you to dance—he said moves only. I thought it was okay with you."

"Well, it's not okay." I got in the Rilken's slimy face. "His moves can kill anything that breathes, you brain wipe."

"They say you're faster."

I should have called the whole thing off right there. Even on Terra, blade dancers had notorious reps—they were all trained to kill anything alive, in so many ways that the Shadow probably didn't have room left in his skull to remember the "no weapons" deal we'd made.

I knew more about blade dancers than the average Terran, mostly from rumors and bogeyman stories passed around the underground. Why hadn't I recognized his type right off? Why hadn't the things he'd said to me clued me in? Like *you move well.*

In comparison to his other victims, I was sure I did.

Someone harrumphed. I looked at the crowd that had gathered around my corner. Eager faces watched me with the intensity of high rollers guaranteed a fixed bout.

Was it fixed?

The Rilken extended a timid coil and patted the back of my hand. "You can win, Terran." To the others he said, "You've seen her move. She can beat him, can't she?"

The crowd erupted with voices in agreement.

"Shut the fuck up." I pushed the Rilken through the ropes back into the quad. For a minute I looked at the door panel, then the crowd. Possible death, or Ichthora. I really had to think about it.

"All right," I said. "I'll fight him."

Everyone cheered, like I'd already won.

"I've lost my mind," I muttered under my breath as I went through my pregame exercises, stretching and toning my muscles for battle. "If I'm lucky he won't kill me in the first two minutes."

I looked up at my opponent, who was not doing anything but staring back at me.

You not belong here.

Damn right I didn't. But somehow I had to get my mind set for this thing, or I *would* end up dead. What would my offcoach scream at me to get me pumped for World Game?

He means to win—has to win, badass blade dancer. Only, you're better, Rask! You've got to be better! You will *be better! Get out there and stomp his goddamn head in the turf!*

Rijor's voice, my voice of reason, chimed in: *He'll try to control the fight, Jory. He's preconditioned by drill and experience, just like us.*

Look at him! my offcoach screamed. *Standing there like he's waiting for a glidebus! Are you going to take this shit from a clown carrying a stick?*

I knew a lot about psych jobs, too. Playing shockball was really two-thirds muscle, one-third brain. If your opponent thought you were the fastest Terran on the planet, it weighed his legs down without his even realizing it. How else could I have outrun every runblock, even when I'd played injured?

Rijor's ghost laughed at me. *The Jorenian blood didn't hurt, either, Jory.*

Shut up, Rij. I climbed into the quad. *Go ahead, pal. Take your best shot.*

The Rilken stepped between us and held up a couple of slimy coils. "Match between the breed and the dancer, moves only. Pin for a ten-count to win."

There were a lot of ways and places I could be pinned. "Shoulders to the quad floor."

"Shoulders to the quad floor for ten to win," he added as he scurried out of the way. "Begin!"

The dancer went still for a moment, then fluidly sidestepped, going for my left side. I went into a block-break crouch and backed away.

The crew began shouting out encouragement, mostly for me—evidently I was the home team. Or the underdog. I kept my eyes on his hands and mirrored the dancer's movements, both of us revolving around the center.

He was so quiet I couldn't hear him move—no dimsilk rustle, no footsteps, no breathing. He never stopped moving, either. Sweat beaded on my brow, and I felt that first, irrational wave of terror slam over me.

He's going to kill me.

Rijor had taught me how to channel my fear, let it block out everything but what was important—scoring.

Use it for focus, Jory. Don't center and fixate; see the entire field. Move; don't freeze up. You should be afraid; everyone around you wants you dead. Make it work for you.

My offcoach didn't mince words. *Quit whining and kick his ass.*

I moved, keeping my fists up and my weight balanced on both feet. The dancer changed his languid pace and strode up to meet me as if he were walking down a corridor.

I struck, then recoiled in shock as he neatly dodged my right.

God, he's fast.

I circled, still countering his position, though we were only two feet apart now. Keeping my left in to protect my head and abdomen, I threw a few more jabs at him. He kept dodging, gracefully turning to the right, or back, just enough to avoid the blow.

We could do this all day, I thought, then had to whirl to the left as he struck back. The blow glanced off the side of my right arm, enough for me to feel the power behind the punch.

Fast and *strong.*

A surge of adrenaline made me slightly dizzy, and I backed off, trying to shake the rush. Now was not the time to get rookie jitters. There was a lull in the noise, and out of the corner of my eye I saw heads turn and heard whispers hiss.

So he swiped me. Not like I'm going to cry.

Yet that one punch made the difference, and the audience began cheering the dancer. As he advanced, some of them even shouted out suggestions on how to pound me into the deck.

Hyenas.

Now I had to dodge him as he came after me. It was like being in a rain shower and trying not to get wet. Still, I managed to keep him from landing a solid blow.

"So this is why you call it dancing," I said, ducking under a punch and trying to get at him from behind.

He whirled and eluded my sneak attack. "You like?"

That was the weird part. I did like it—a lot. Fighting felt like shockball, only I had to go for the score *and* defend myself, instead of relying on blockbacks to protect me.

I'd nearly socked him in the abdomen when something totally bizarre happened. My vision blurred and my stomach knotted. I'd never felt anything like it. Then I did something I'd never done in all my years of playing pro ball, even when I ran injured.

I tripped over my own feet.

He should have jumped on me then, but incredibly, the dancer hesitated. I latched onto the ryata rope and fought for balance. The stick appeared in his left hand.

The dizziness was bad enough, but this? "Hey! No weapons!"

A blade shot out of the end of the stick, unlike the other two he'd pulled before. This one looked shorter, sharper. The stick dwindled and darkened, and became a blunt-looking hilt about a split second before he threw it at me.

Really stupid idea, Jory. I braced myself for the impact, but the blade flew past my belly, missing me completely.

Someone screamed; then things got very, very quiet.

I hung on to the ropes and turned to see one of the crew lying half in, half out of the quad. It was the friendly, stalk-eyed one who'd slapped me on the back before. His arm was flung out, pinned to the deck by the blade sticking out of it. A strange-looking emitter sat next to his spasming hand.

By then the dancer was standing over the crewman. He picked up the emitter and switched it off.

My nausea and dizziness vanished.

"Brain tumbler." He brought it over and handed it to me, then reached around to pull something off the back of my neck and placed a tiny adhesive chip on my palm. "Tagged you. Signal made you trip."

I studied the device and the chip for a moment, then glanced at the sobbing, cringing crewman. "Hedging his stake."

"Yes." The assassin returned to the downed man, bent and jerked the blade out of his arm, then did something to make it resume the bland stick form. "You kill him."

I tossed the device between my hands as I considered it. "I'd like to. Slowly."

He removed another, shorter dagger from inside his garment, making me wonder just how many knives he had hidden on him. "Use this."

"I don't need it." I flexed a hand, flashed some claw.

That impressed the crowd. "Suns, she's Jorenian."

"I thought they were blue," someone said. "She's pink."

Another made a disgusted sound. "She's a breed, you thick skull. They come in different colors."

It felt good to see the crewman's eye stalks stiffen and his eye clusters bulge just a little bit more, but I retracted my claws. "Not worth the hassle."

The assassin gazed down at the crewman. "Agreed."

The Rilken who had arranged the fight edged up to us. "So? Will you continue?"

The fight felt like a stalemate. I couldn't beat the dancer's agility; he couldn't beat my speed. There's always a point when you have to accept that something's pointless.

"No, the fun's over. Unless this idiot"—I nudged the terrified crewman with my foot—"wants to dance with one of us."

My opponent looked at the Rilken. "You arrange this?"

Now that wasn't something I'd considered, but maybe my brains were still a little scrambled.

"No, no, I knew nothing about it." The little alien paled as the assassin fingered his blade. "I swear to you, that's the truth."

"It had better be." I crushed the chip under my boot before I tossed the emitter to him. My leg throbbed, and my stomach still felt queasy, so I decided to head to my cabin. "See you later, boys."

The dancer took a bulging credit holder from the Rilken and slapped it into my hand. "For trouble."

I didn't like charity, but I wasn't going to argue with him. I curled my fingers over the holder. "Thanks."

I strode out of the quad, making sure not to limp until I reached the corridor. Yet even after the door panel closed behind me, I could still feel the assassin watching me go.

CHAPTER FOUR

"THE WISE TRAVELER KNOWS HIS DESTINATION."

—TAREK VARENA, CLANJOREN

After checking my knee, which despite the stress I'd put on it looked fine, I went to my console to check on the database search I'd started that morning. The display showed the results:

> List any/all news/immigration/bonding/civil/criminal/penal/obituary files on native Jorenian citizens born ****28:****29 with following names: Jakol Varena. Nalek Zamlon. Galena Nerea. Osrea Levka. Danea Koralko. Renor Xado.
>
> Database search complete. No files found.

Which meant they were all still citizens of Joren, not bonded, not in prison, not dead, and hopefully on-planet. As for my second search string—*list any/all news/immigration/bonding/civil/criminal/penal/obituary files on native Terran male surname Kieran*—the screen offered one file found:

> Found one (1) on native Terran male Kieran/cross reference: order for apprehension and detention.

I opened the file and began to read.

> Allied League of Worlds Transport Security Apprehension and Detention Order: Planetary and Trade Route Authorities are advised to immediately apprehend and detain the following individual, Terran male,

PID63915179, known as Kieran, wanted on multiple counts murder, theft, piracy, assault, unauthorized detention, destruction of property . . .

The list of Kieran's offenses was further broken down by quadrant, the details of which continued for two more screens. The file ended with:

Last reported physical description of subject: Biped humanoid, 100.23 kilos weight, 1.824 meters height, standard Caucasian derma, black hair (upper cranial case only), green eyes (two) . . . no photoscan available.

Obviously Kieran didn't like having his picture taken, I thought, and input a third request:

List any/all news/transport/recovery/demolition files on Jorenian Star Vessel MoonWave.

The console bleeped an acknowledgment, and after several minutes produced one more file, a multiple-quadrant alert that had been transmitted and archived twenty-seven years ago.

Located: database STS relay archive file ALW/TW16914107

I rubbed my palms on my trousers before opening the file.

ALW ITS relay 100327/pridistr/poTW725426: Allied League of Worlds Interstellar Transport Security requests any/all data pertaining to the current location and disposition of Jorenian star vessel MoonWave, TWSID M2991E5070V. Vessel's last transmission indicates possibility of raider pursuit and/or capture. Transmit data to ALW ITS via SGRC8354145 . . .

No mention of Kieran, or the others. So Mom was right; Joren hadn't reported any of the specifics regarding the *MoonWave* incident back to the Allied League.

Why not?

From what Mom had told me about her people, they didn't like anyone messing with their blood relatives. Someone from her HouseClan and all the others involved had undoubtedly copped the right to track down Kieran for what he'd done to the *MoonWave*'s crew. But big as the galaxy was, how could they hope to find one cold-blooded mercenary who didn't want to be found?

I looked at the screen and saw I had typed *BLADE DANCER*. My database had interpreted that as a query, and offered a generic definition:

> Blade Dancer: proper noun [blAd 'dan(t)s&r] Etymology: Archaic Terran blœd + dancier. One of a covert order of assassins for hire, one who murders with a transmogrifiable weapon (see tån). Origin of order: Planet Reytalon, charted location unavailable.

I put in more queries. Outside of the unilang dictionary, I found very little hard data on the dancers. Some sensational stories that read like bad League propaganda, a few speculative theories from news relays, and a list of planets that refused to grant them immigration visas.

A long, long list.

Not that I imagined any blade dancer would covet residential status. An assassin's profession meant plenty of travel, as well as the need for anonymity. Blade dancers had to be homeless, constantly on the move, always pursuing their target. I could even sympathize—I certainly knew how it felt to be forced to keep moving and never let anyone know who I really was.

I switched off the terminal. Over the past revolution, blade dancers had been blamed for more than a hundred assassinations in seven different quadrants, but not a single one of the elite killers had ever been seen or identified, much less caught.

They might be scary, and murderers, but you had to admire that kind of skill. I might make a half-decent blade dancer, myself.

You're nothing like them. My conscience sounded a little worried. *You've only got to get to Joren, and you'll have a real family, and a home, and the chance for a new life.*

The problem was, I couldn't see myself setting up shop on Mom's homeworld, or trying to fit into a culture I'd only heard stories about. 'Gill's warning about my knee ruled out contact sports, if they even had them. But once I'd kept my promise and contacted the others, what could I do?

I'd never been trained to be anything but a runback who could pass as Terran. No one was going to hire me because I used to run fast and knew how to spit like the best of the xenophobes.

They won't be like the Terrans. They'll help me make a new start.

I lay on my sleeping platform and stared at the upper deck for a few minutes. The Shadow had been so hot to challenge me, would have beaten me—I could admit that now—and had ended up defending me. Not exactly the kind of thing you'd expect from someone who made death his living.

Why would a killer want a fair fight?

I drifted into sleep, and dreamed of fighting dozens of dancers, all of them trying to carve my eyes out with their blades. I woke up thinking I was screaming, then heard the screech coming from my panel.

"—repeat, all crewmen to defense stations!"

Defense stations meant someone was attacking the ship. Not a very bright someone, either. Gun runners like the *Chraeser* maintained state-of-the-art firepower. Unless it was someone who had better.

Since I had no intention of waiting to die in my cabin to find out, I grabbed my knife and headed out to the corridor.

A crewman distributing pulse rifles from a storage unit at the other end of my level tossed one to me. "You want another fight, breed, head down to cargo level nine."

I checked the power cells, which were full. Something hit the port side of the ship hard enough to send a shudder through the entire hull. "Who am I shooting at?"

"The Hsktskt."

Ah, the reptilian butchers who either killed you or enslaved you, currently at war with anything humanoid. My kind of target. "Any League ships around?"

"What do I look like, an S.O.?" He tossed me a spare charged cell. "Go kill something green!"

I was stopped the minute I stepped off the lift onto nine, but a security guard recognized me from the fight and let me pass. Temporary barricades had been set up to provide cover, but every hold on the level had been crammed full with stockpiles of League weaponry and armaments. If anyone got into serious shooting down here, it would all be over real fast.

There were worse ways to go. Word had it that the Hsktskt liked their food still warm and verbal.

Most of the crew had bunched up behind the barricades, but I wanted a better vantage point. When I'd pulled watch for the underground, I'd climbed trees—and remembering that made me look up. The level's ceiling wasn't all solid deck; there were some maintenance-access hatches at regularly spaced intervals.

At the farthest end of the level, I climbed an access ladder and pushed open a hatch. That led to a maintenance passage, but it was too short for anyone but a Rilken to walk upright through it. I backed up into the crawl space, propping myself over the open hatch. Not much room to move, but better concealment than the barricades below.

As I waited, I felt the ship take more displacer hits. The optic emitters flickered as the hull absorbed the impacts. Someone uplevel must have cut back power to the envirocontrols, because the temperature began to drop.

No, they're doing it deliberately. The lizards don't like the cold.

Another enormous blast shook the lower deck, and someone cursed. "*U'flargot* scum, got the main cell. They'll be boarding through launch bay, level six. Disable the lifts!"

Security shot out the lift controls and locked down the doors manually. Everyone stopped moving around and got very quiet after that. Occasionally I heard a murmur, like someone praying.

Not a bad idea.

"Mother of All Houses," I said, watching my breath appear in little white puffs in the freezing air, "You never listened to my Mom, did You? Now would be a great time to make up for that and save my ass."

More displacer shots exploded, but these felt closer—inside the ship. The guards abandoned their posts in the open corridor and took up positions behind the barricades. I activated my rifle and aimed for the lift doors.

"*Uhsstaaa,*" something hissed behind me. A hard, scaly hand seized my ankle and started dragging me back into the crawl space.

Hsktskt.

I dropped the rifle, which fell to the deck below, then grabbed the edge of the hatch with both hands. I couldn't let go to get my knife. If I yelled, everyone below would shoot up at me. The alloy threads on the hatch opening bit into my fingers as I fought to hold on and keep silent, but someone was already climbing up the access ladder.

The blade dancer.

The Shadow looked up at me as he drew out a sword and a smaller blade, then made a gesture for me to put my head down.

If he decapitates me trying to stick this Hsktskt, I thought, *let me stay alive long enough to stab him in the heart.*

I pressed myself against the bottom of the crawl space. A moment later the dancer popped up through the access hatch, swinging the sword down and throwing the dagger.

"*Nsseerok!*"

I felt the whoosh of air against my leg, then heard the hissing voice turn to a sloshy gurgle as something hot and wet splashed across the back of my trousers. Whatever had me by the ankle went limp.

I craned my head around to see the Hsktskt raider, his clawed hand clutching the dagger stuck between his eyes. One of his limbs lay severed beside me.

I let go of the edge and slumped against the cold metal under me. My heart was pounding so hard it should have dented the alloy. "Jesus Christ."

"More soon." He straddled me, but only to reach back and retrieve his weapons. I heard a disgusting sucking sound and felt the Hsktskt's body twitch as the blade slid out of his skull. "Down now."

I followed him back down the ladder, expecting him to shove me behind a barricade. Instead he pointed to a large open section. "In there."

"There's no cover," I said as I retrieved my rifle and watched the hatch.

He handed me the dagger. "More room."

"No, thanks." I handed it back. "I've got my own."

"Then use it."

An explosion blew out the entire lift section, and the dancer hauled me into the open hold. Shouts erupted, along with pulse and displacer fire. He took position on one side of the door panel and pointed at the other.

"Ready?"

No, I wasn't ready. I should have left him there and found a nice, strong barricade to hide behind and fire from. But I owed him my life, and for some reason I suspected staying close to him improved my chances for hanging on to it, too.

"Yeah." I dropped the rifle and pulled out my knife.

"Omorr?" he murmured, and I nodded. "Good blade."

For a moment I wondered if he was crazy—we were facing the Hsktskt, my stomach had wedged itself in the back of my throat, and we were almost certainly going to die. And he was admiring my knife. "Thanks."

Heavy footsteps pounded down the corridor, coming closer with every second. We heard the barricades fall as they took them out, one by one. The sounds of agonized screaming, ripping flesh, and breaking bones made me close my eyes for a moment.

You should be afraid; everyone around you wants you dead. Rijor's ghost chuckled. *Make it work for you.*

"Forget fear. Forget everything."

I looked across the open doorway. "What are you, telepathic too?"

"Keep close to them. Cut big tendon on back lower limb, above joint. They face you, slash eyes. They go down, cut throat. Keep moving."

He was telling me how to mutilate Hsktskt. "You've killed a lizard before?"

"Many."

I tensed as the first two raiders lumbered in through the door panel. *Now I live or die.*

Live, Jory, my mother's ghost whispered. *Live for both of us.*

I didn't want to die, so I went for the tendon on the one closest to me. The dancer did the same. Impact sent a shock wave up my arm, but the Omorr blade sliced easily through the lizard's tough scales and ropy muscle. I blocked out the roars and stepped out of the way as he fell to the deck. The dancer kicked their weapons across the hold, and we cut their throats in tandem.

It was disgusting, seeing their flesh open, watching the blood spurt. Knowing I was responsible. I wanted to vomit.

Forget everything.

More Hsktskt charged in, and suddenly I became too busy hacking at scaly faces and legs to notice what my mentor did.

Keep close to them.

Displacer fire echoed wildly all around me, and the big reptiles kept body-slamming me, but I fought for balance and chanted the dancer's instructions in my head: *Cut the tendons. Slash the eyes. Cut their throats. Keep moving.* And as survivors from the corridor burst into the hold, I added one of my own: *Don't kill anyone that looks like crew.*

One of the raiders hit me with his tail, sending me sprawling, but I'd taken too many tackles on the field to stay down. I used the momentum to hurtle back up into a crouching position, then cut the lizard's feet out from under him as he came at me.

"The tendons," the dancer said as he thrust his sword in the Hsktskt's throat. "Not the feet."

I blinded another Hsktskt, then sidestepped as he went down, roaring and clutching at his face. "Critic."

The skirmish ended several minutes later, when no more Hsktskt charged the hold. The few surviving crew members backed up against the wall panels, panting and holding whatever parts of their bodies were bleeding.

"God." I felt battered and exhausted, and my arm wanted to wither and fall off my body. "Is that it?"

"That it."

I swiveled to see the dancer walking from lizard to lizard, pinning the live ones with a foot before slitting their throats. Although it made me want to vomit again, I made myself watch the way he did it.

When he was finished, he came to me. "You never killed Hsktskt before."

I'd never killed before, period. "No."

"You use blade before."

That I had. Since I owed him, I offered him my gore-covered knife. "Here. For keeping me alive."

"Blade keeps you alive," he said, then turned and walked out of the hold.

I looked around me, then ran to the nearest disposal unit, and gratefully emptied my stomach in it.

As it turned out, ambient temperature defeated the Hsktskt raid to take over the *Chraeser* as much as our battle on level nine. A timely SOS also summoned a large squadron of League strafers, who showed up to polish off the lizards' attack vessel, then flew escort for us for the rest of the jaunt.

While the crew joked about new ways to turn the ship into an instant blast freezer, the captain summoned me to command.

"It was self-defense," I told him as I came in. "I have witnesses."

"I heard. Your passenger fee is refunded to you, and we will transport you anywhere you like." He pulled up a star chart of the surrounding systems. "Pick a world."

I didn't think my minor contribution to the battle on level

nine rated that kind of reward. "Why the sudden burst of generosity?"

He seemed surprised, then smirked a little. "Ask the dancer."

I tucked my hands in my trouser pockets. "I'm asking you."

"This ship belongs to him, Terran. He orders; I obey." He shut down the holoimage. "He's interested in you."

"Is he." I rubbed one of the strained muscles in my arm. A blade dancer didn't get interested in people. He killed them. "I'll need to talk to him first."

"I'll pass the word along. He'll find you."

Finding me wasn't difficult. I went to the galley and sat until he appeared beside my table. He gestured for me to follow him, and we walked together down the corridor. Crew members nodded to us, their expressions ranging from fearful to near-worship.

I guess everyone knows he owns the ship now, I thought as I watched them. *Or they heard how well he can kill.*

He stopped at a heavily secured door panel and pressed his gloved hand over an access port.

"What's this?"

"My quarters."

The door slid open, and while I debated my sanity in doing so, I followed him in. There wasn't much to look at inside—sparse furnishings, a single table and chair, a prep unit, and an equally well secured storage unit.

"Tell me, is this some kind of weird assassin type of courtship thing?" I said as I went to his viewport. "Because if it is—"

"It is not."

"Great. Then what is it?" I turned around. He was standing in the center of the room, completely still—he would have made an exceptional statue. "What do you want from me?"

"Nothing."

I seriously doubted that. "You just refunded my passage because you felt like it?"

"My gratitude."

"Bullshit." I waited, but he didn't respond. Maybe my wristcom hadn't translated it. "What if I want to go to Joren?"

"Joren not for you. Go to Reytalon."

So he *was* some kind of killer recruiter. "To train to be like you?" I walked a circle around him. "Why would I want to do that?"

"Revenge takes practice." He drew out a dagger. "Like killing."

Even though it made no sense, I didn't feel afraid of him anymore. "You'd know." A thought occurred to me. "That's why you wanted to fight me. It wasn't practice. It was an audition."

"Yes."

This was going beyond spooking me now, but I kept up the illusion of indifference. "I don't want to kill for a living. It makes me sick." But the idea made me think. "If I went to Reytalon, could I learn to handle blades the way you do?"

"You must go Tåna."

I frowned. "Is that like, what, assassin school?"

"It can be."

It was almost funny to think there was a school where you could learn to be the most lethal assassin in the galaxy. Then everything seemed to fall into place at once. "How do I get in?"

I expected some kind of secret rendezvous point or complicated verbal instructions. Instead the dancer went to the room console, pulled up a star chart, and highlighted a planet. "This Reytalon."

I went over and studied it. If the dot he was pointing to was Reytalon, it was in an uncharted system far beyond Joren. I memorized the coordinates. "They take females?"

"They take anyone." He erased the chart. "They teach you *shahada*, dance with tån."

"Then what happens?"

"You kill." He watched me as I started pacing. "You want kill."

Was he trying to convince me, or talk me out of it? "What if I only want to kill one man?"

"Who?"

I couldn't see explaining the whole mess, or arguing the point. "His name is Kieran. He's a mercenary and a raider for hire." Among other things.

"Kieran blade dancer. Dangerous man."

I stopped pacing. My *father* was a blade dancer? "You know him?"

"Who does not?" The dancer tapped the screen. "Speak to Tåna Blade Master. He know Kieran."

Plans began forming in my head. "I can't go to Reytalon yet. I have to go to Joren first."

"Captain take you Joren."

I went to the door panel. "I owe you for this, dancer."

"You kill Hsktskt, Terran," he reminded me.

I *had* taken down a lot of lizards. Without his help, it might take months to reach my mother's homeworld. I glanced back at him. "All right. This makes us even."

The meet with the dancer had given me a lot to think about. Why would anyone want to recruit me to be an assassin? I was an athlete, not a killer. Until the Hsktskt had attacked us, anyway. Then I found myself wondering just how hard it would be to get through the training, and what the Tåna would expect in return. If this Blade Master knew Kieran, he might give me a lead on how to find him. And then—

"I hear you're going all the way to Joren," one of the League soldiers said as he caught up with me in the corridor.

"Maybe." I recognized him as the one-eyed corporal who'd spoken to me on the training deck a few times. "You?"

"I'm for Vgfria Station, near the Varallan border." He gestured for me to go ahead of him into the training room. "You want me to spot you on the free weights?"

I usually resorted to a drone spotter, but having some real company would keep me from dwelling on my mysterious benefactor.

"I can't get over that extension unit buckling the way it did," he mentioned as I stretched out on my back and he took position by the bar supports. "You ever think about how close that was?"

"I'm not going to forget nearly being crushed to death."

He huffed out a laugh. "Someone wanted you dead in a big way."

"That's another thing." My brows drew together as I gripped the bar and found the proper balance spots. "Why would someone try to kill me?"

"Don't know. Maybe someone just wanted to see how fast you are." He watched as I lifted, lowered, then lifted the weights. "One-fifty, not bad. How many reps do you pull?"

"Twenty." I dropped the bar back on the supports and sat up. "Excuse me."

"Hey, you've got nineteen to go!"

I hit the access panel and strode out into the corridor, nearly knocking over a wide-bodied Dakrith.

"Watch where you're going," he snarled, then stumbled back as I turned on him. "Sorry. My fault. Won't happen again."

I didn't storm into the dancer's quarters. I signaled politely. I could be civilized about this. I could find out what his agenda was. The door panel opened.

Or not.

I lunged, knocking his stick out of his hand and pinning him against the nearest wall panel. "You sabotaged the weight unit. Admit it."

"You have speed. Strength. Courage." He took out another stick and transmuted it into a blade, then rested the tip against my belly. "Need wisdom."

"Obviously." There was nowhere for either of us to go, so we were staying right here. "Why all the games? Why not just kill me the first time I stepped on the ship?"

"I never play." He moved then, somehow sliding out of my strong hands, whirling his blade up in front of my face. I jerked back, but it was so close I could feel the air displaced by the edge whisper across my skin.

I backed into the corridor, watching him advance, still holding the blade. "You're a blade dancer. I know what you are. Why am I alive?"

"Because you are fast." He brought his knife to within a centimeter of my right eye. If I blinked, he'd slice open my eyelid. I could smell a strange, pungent herb on his breath as he leaned in close, still not touching me anywhere. "You go Reytalon."

"Why?"

He grabbed me then, whipping me around until my face collided with the wall.

"Go to Reytalon," he murmured next to my ear. "Find Kieran. Kill him."

I was going to Joren and forget about this whole bizarre situation. But he had a blade, and obviously expectations, so I lied. "I will."

His grip vanished. I tensed, but all I felt was another whisper of air against my neck.

When I turned around, he was gone.

CHAPTER FIVE

"BEWARE THE OBSTACLE THAT SEEKS THE PATH."

—TAREK VARENA, CLANJOREN

Joren wasn't exactly what I'd expected. Oh, it was just like the planetary survey data said—the seventh planet in a single-star system, big, multilife-form sustaining, etc. The moment I stepped off the launch at Lno Main Transport, though, I knew they'd missed a few things.

I breathed in deeply. "Nice. I could dab the whole place behind my ears."

The air smelled like flowers. Colors were sharper, clearer. Even the gravity felt right—on Terra, I'd always felt like an oversize, lumbering *t'lerue*. Now I felt my muscles shifting, as if finally relaxing after twenty-five years of tension.

Amazing what an extra couple pounds of atmospheric pressure per square inch would do for you.

As I walked down the ramp, a big blue-skinned female walked out of a nearby building and approached the launch. She was a little shorter than me, but wore a fancy green tunic with slashes of purple and blue on the sleeves. I'd given my mother a reproduction Tiffany lamp once with the exact same colors in the stained-plas shade, and she'd kept it by her bed until the day she died.

My spirits took an abrupt dive. Everything on this damn planet was probably going to remind me of Mom.

Why should it not? her ghost wanted to know. *This was my home.*

The Jorenian female didn't stop at the ramp, but strolled up to meet me. *Transport Admin*, I read from the discreet badge on her chest. A well-paying job, judging from the amount of expensive body ornaments she had hanging from her ears, throat, and wrists. She gave me a not-too-enthusiastic smile and reached up to put something around my neck.

I ducked out of the way. "No, thanks."

"Loceg Jorenhai?"

"I speak Jorenian," I said in a coherent, Terran-accented version of the same.

"Forgive me. I assumed as a human you did not, and only wished to give you a vocollar to ease your path." She indicated the necklace of translinks.

Interesting, the way she said *human*. The same way I'd say *dirtbag*.

Be nice, Jory, my mother's ghost said.

I could do nice. I shifted my case from one hand to the other. "Thank you, but I think I can make myself understood."

"Very well." Her expression remained polite but clearly indifferent. "I am Enale Raska, visitor guide. May I assist you in reaching your destination?"

Raska. That explained why she was wearing Mom's HouseClan colors—she was a relative. Might as well make that my first stop. How did Mom tell me to say it? "I bring news of some importance to relay to your HouseClan, ClanDaughter Raska."

That thawed her a little, and she smiled. "You are most welcome, then. Come, I will guide you there." She paused at the end of the ramp. "May I know your name, lady?"

So I was welcome, but only because I was the delivery girl. "Call me Jory."

Enale didn't have much to say as she drove me from Main Transport outside the city to HouseClan Raska's territory. I mostly blocked her out and drank in the sights I'd only heard Mom describe to me in stories.

Yiborra grass fields, stretching in pools of silver in every direction. Bloodred jaspfayen flowers, their chambered petals singing in the breeze. Thin streamers of cloud overhead, in a thousand rainbow colors, echoed on the rippled surfaces of a placid, purple-water lake. Half the population still dwelled out here, on huge tracts of HouseClan land in communal pavilions with their entire family.

Joren is like no place in the universe, my ClanDaughter, Mom would say. She'd tell me all about it in a fond, remembering voice. Then, when she thought I was sleeping, she would weep for hours.

"Is this your first sojourn to Joren?" Enale asked as we pulled up outside a huge, sprawling structure built of white stone and etched with pictographs in all shades of the purple, blue, and green Raska colors.

"Yes." I thought of Reytalon and the blade dancer. He was wrong about me. I could stay here; I could make a life for myself and forget all about Kieran. I didn't even have to find the others. *I can do this. I can start over here.*

"If you have the opportunity, you should visit Gafa Lno Lake country before you leave. It is quite beautiful this season." Enale glanced at me. "May I remain to hear the news you bring, or is it a matter of confidentiality?"

I wasn't sure how the Raska were going to react, but I couldn't see confiding in this ice princess. *Before you leave—now that's subtle.* "I'll have to check with your ClanLeader."

Mom had told me about the pavilions that served as home to the HouseClan's leader and his immediate family, and as a meeting place for the rest of the kin. I tried to imagine Terrans living and working together with their extended families.

They have a hard time tolerating each other for the holidays.

The interior impressed me more than the outside of the place. Expensive furnishings in the free-form, no-corners style I'd seen at Main Transport were artfully arranged to provide little conversation clusters, beneath artwork in the same green/purple/blue colors. Jorenians seemed to like a lot of landscapes and star vistas, but there was one portrait of a stern-looking, big man with a slightly smaller, equally stern-looking woman.

"Who are they?" I asked Enale, nodding to the portrait.

"Our ClanLeader, Skalea Raska, and his bondmate, Tnefa

Raska." She made a peculiar gesture, then explained, "My ClanMother Tnefa embraced the stars during the last rain cycle. Those who journey beyond us are celebrated with delight, but I . . . miss her."

So the ice princess did have some warm blood in her veins, and, although she didn't know it, she was my aunt.

Be nice, Jory.

"I lost my own mother a few weeks ago," I heard myself say.

Enale looked uncertain. "I am told offworlders do not find joy in the path that is diverted."

"No, death tends to be a little inconvenient." When that confused her even more, I shook my head. "Never mind."

She smiled briefly, then escorted me over to some kind of main console, where she signaled someone. A happy voice directed us to the "ceremonial chamber," where everyone was working on "preparations for the feast."

"Having a party?"

Enale nodded. "Tonight begins *Nadamar*, the sacred time of the Mother."

The ceremonial chamber was huge—much bigger than I'd expected. It could comfortably hold over a thousand people, and from the food being laid out they could feed twice that many. As we passed members of Enale's family, I noted how everyone appeared to be uniformly attractive—evidently there was no such thing as an ugly Jorenian—and happy. The smiles and laughter were plentiful and practically nonstop.

If I had to live here, I'd want to punch someone out by the end of the first week. Two weeks, tops.

Still, the healthy-looking males and females seemed to be in the perfect mood for a party. It fascinated me how even their smallest movements seemed graceful and effortless. They spoke in melodic voices and gestured with fluid hands, especially with their children, who were evidently the cleanest and most well-behaved kids in the universe.

Poor Mom. She expected one of these, and got me instead.

I could pick out older members of the HouseClan only by the vivid streaks of purple in their matte-black hair. No stoop-shouldered, wrinkle-faced senior citizens on this world. A couple of males with solid purple hair lifted a heavy-looking table and shifted it across the room with every indication of ease, so I guessed advanced age didn't equal infirmity here, either.

As for the banquet, the theme was flowers and more flowers. I'd never seen so many in my life. Every color of the rainbow, they hung from the ceiling on ribbons, sprouted from enormous open silver baskets, and adorned countless platters of fruits, vegetables, and bread. They smelled even better than they looked.

"You are welcome to stay and join our celebration," Enale added when she noticed my interest. I imagined she'd invite a Hsktskt with exactly the same lack of enthusiasm.

"Thanks." I was watching one of the solid purple-haired Jorenian males head our way. He wore a different tunic than the others, and had a funny, knotted thing that hung from his shoulder and went around his waist. The closer he got, the less he smiled. "Here comes your ClanLeader."

"ClanFather!" When the ClanLeader halted and made a formal gesture of greeting, Enale returned it with a much bigger smile and fluid hands. "ClanFather, this is Jory, a visitor who just arrived at Lno Transport. Jory, this is our ClanLeader and my parent, Skalea Raska."

My ClanFather is such a wonderful man, Mom had told me once. *I wish things had been different. I know he would find so much delight in you.*

I made a gesture of deep respect—awkwardly, since I'd only practiced it a few times with Mom—but completely sincere. "Greetings, ClanLeader Raska."

Skalea's white-within-white eyes inspected me like a line judge on third down. "ClanDaughter, excuse us for a moment."

He didn't like me any more than the ice princess did. I could feel it. When Enale departed, he stepped a little closer. We were exactly the same height, but he couldn't see through my dark lenses.

"Your kind are not welcome here." He slashed his hand back toward the entrance. "Leave our House."

I'd expected them to have a little dismay over the color of my skin, but still, after showing him the proper respect it was like being slapped in the face.

And I didn't like getting slapped. "Don't you want to hear my news first, ClanLeader?" I let my lip curl on the final word.

A couple of Raska nearby stopped what they were doing to stare at us. He put his hand on the funny belt part of the thing he wore, and I saw the dull gleam of a knife hilt appear above his fist. "Say what you will; then go."

"Kalea Raska is dead." Since I'd never been one to crash a party, I turned on my heel and headed for the exit.

Skalea caught up with me and put his hand on my arm. To an onlooker it might have appeared like a charming, old-fashioned gesture. They couldn't feel the grip he had on me. "You will come with me now."

I didn't like Jorenian hands on me any more than I liked Terran, so I shook him off. "Uh, no, I won't."

"Kalea was my eldest ClanDaughter. I would know how she embraced the stars." He pointed to a corridor off the banquet hall. "If you would accompany me to my chamber, there." He studied my face, then added, "Please."

That last word cost him, so I went along to his office. Once inside, he secured the door panel and gestured for me to sit down in one of the comfortable-looking chairs in front of a wide, U-shaped desk.

I didn't need his grudging hospitality, so I stayed on my feet.

His eyes narrowed as he sat down behind the desk, but he didn't make a fuss about looking up at me. "Kalea embraced the stars from Terra?"

"Yep."

"How did her time come to her?"

I saw no reason to spare him the gory details. "She contracted a Terran virus lethal to Jorenians, was unable to seek medical treatment for fear of deportation, and died alone, in pain, screaming." I smiled politely. "Get the picture?"

He sat back and stared past me for a few seconds, his expression blank. Then he made a gesture of acceptance. "She chose her path."

Oh, so it was okay with him? My fantasy about making a fresh start on Joren collapsed into a heap of torched dream rubble.

"After your ClanDaughter died, they put her into a disposal unit and reduced her body to recyclable organic matter." I came forward, leaned over his desk. "I can guarantee you, ClanLeader, she didn't choose *any* of that."

Skalea gave me a suspicious frown. "How did you become involved with Kalea?"

He hadn't guessed who I was. Gee, I might actually have a little fun with this. "The authorities found me trying to bury her body in the desert. They deported me for violating their laws."

"That is not what we do with our kin." His hands clenched. "Did she Speak to you?"

Now I lied. "No." My plans in ruins—a lot like what was left of my heart—I decided it was time to go. "Any other questions?"

"Who are you? Why did you meddle in my ClanDaughter's affairs?"

Meddle. That was the final straw. "Can't you see the resemblance?" Now I took off my shades, then my gloves, and tossed them on his desk. "I'm not as tall, and the eye and skin color came from Daddy—but I have her look, don't I?"

Dark color rose in swatches over his angular cheekbones, while the rest of his face turned a powdery blue. He made a gesture, unable to speak.

I moved in so he could really choke on it. "She named me Sajora, by the way. After your ClanMother, right? But call me Jory." I braced my hands and got close enough to kiss the end of his big blue nose. "And what do I get to call you? *Grandpa*?"

"You are not of this House!"

"Oh, boo-hoo, now you've hurt my feelings." I pouted. "I'll have to cross you off my Christmas card list."

"You—you—" He struggled up from the chair, then sagged a little. His pale face looked haggard, and he wouldn't meet my eyes. "Why did you come here? What do you want from us?"

"To tell you that she's dead. That's all." I covered up my eyes and hands again. "Good-bye, ClanLeader."

"Wait." He made a small gesture of apology. "Your pardon, I had not expected . . . we never knew if her child had lived."

"Well, I'm her child, and now you do." I didn't want to ask about my father, but I had to know. "Have you killed the man who sired me?"

"No. The oath we took prohibited seeking retribution against him and the others. There is much to discuss." He reached the door panel before me, and went on, though his lips were white and every word seemed a huge effort. "You may petition the Ruling Council, and if you are sincere they will doubtless grant you residential status. I cannot speak for the Raska until I hold House council, but it is unlikely that my kin will accept a child born outside bond."

Though I felt like I was being spit on again, I interrupted him with a laugh. "What gave you the idea that I would want to live on Joren or have anything to do with you people?"

"I will speak to the councils for you—"

"Fuck you, Grandpa." Out I went.

"Stop." He paced me in the corridor. "You would not journey here simply to relate what you could have by transmission." I glanced sideways at him, and he lifted his head to look down his high-bridged nose at me. "You are Terran; you want something."

Didn't he know when to quit? "You're right. I could have signaled. I did want something." A little payback was in order. "I wanted to see your face when I told you."

"That makes no sense." He got in front of me, and I stopped. "What were Kalea's instructions to you?"

"She didn't have time to instruct me. She was too busy hiding for the last twenty-five years." He didn't blink, and I felt my beast twist inside me. "Do you know how she had to live on Terra? Underground, in caves and tunnels, like an animal? Do you know how much and how long she suffered, just to protect and provide for me?"

"It was none of my doing." He made another gesture of acceptance. "Kalea chose her path—and yours—before you were born."

"You think my mother *chose* to live in fear and die in agony? I don't think so," I said, snarling. "You drove her away because she was pregnant with me. You turned your back on your own kin when she needed you the most. Why don't you announce *that* at the big party tonight, you snotty son of a bitch?" I pushed past him and stalked out to the front of the building.

Enale was waiting for me by the main pavilion entrance. She tried not to look too curious. And failed. "Jory? May I escort you to a guest room?"

Skalea's voice cracked across the room like a slap. "Stay away from this Terran, Enale."

"I might be contagious," I said, to see her jump.

To me he said, "Take the transport, Sajora. It is yours for the duration of your stay, as long as you agree not to return here again."

I studied the transport. It would be handy. "Wild *t'lerue* couldn't drag me back."

"Sajora?" Enale looked at me, then her ClanFather. "I do not understand. That is a Raska name."

"Not anymore, Auntie." I climbed in the transport, slammed on the ignition, and took off.

Three weeks later I found the last of the others, living in unacknowledged exile from his own HouseClan.

The various ClanFests celebrating *Namadar* provided excellent cover for me, which I had needed, for I'd swiftly worn out my welcome on Joren. HouseClan Torin, who were throwing yet another party, seemed like a loud, happy bunch. They had plenty of food, floral wine, and performers all around the ceremonial grounds, putting on song-plays and reciting Tarek. The only spot of calm was the ClanSpar quadrilateral on the east perimeter.

No one went near it.

That's where I found the guy I was looking for. Jakol Varena. The male with the reputation of being the biggest badass in the province, and the only guy on Joren who looked like me, thanks to his half-Terran DNA. He was also the final name on Mom's list.

I used the crowd as much as possible as I approached him, until it thinned out to nothing. The best way would be to come up on his back, get close so I wouldn't have to broadcast my request. A couple of HouseClans, including my mother's, were rumored to be discreetly looking for me, and I had no desire to end up cooling my heels in the Torin's detainment center.

Jakol paced around the five-meter patch of ground, his body close to the ryata roping off the four-sided sparring zone. He reminded me of a Terran tiger in a cage, restlessly looking for a way out. Sneaking up on a defensive liner after third sphere would have been easier.

Turned out it didn't matter. Once I got within a foot of him, I realized he wasn't looking at anything but his own oversize feet.

And damn, he *was* big. A foot taller than me, and easily twice my weight. On top of that, the guy was paved in deeply tanned Terran Caucasian skin, plenty of glistening, sweaty muscle, and attitude. I couldn't smell him, though, so he must have had the same odorless sweat as me. He wore a pair of trousers but nothing above the waist, as if he *wanted* to flaunt his alien hide. Dead-straight black hair streamed over his shoulders and halfway down his broad back. He might have passed for a Jorenian, if not for the skin and his five-fingered hands. From what I could see of his face, he wasn't in a very good mood.

That would work. Neither was I.

"Jakol Varena."

He didn't raise his head, miss a step, or otherwise indicate he'd heard me. Blasted Jorenian restraint. I *needed* to get this wrapped up. The dancer's ship hadn't waited around, and the trader I'd hired to transport me—a nasty piece of work named Uzlac—wouldn't stay in dock forever, no matter how many credits I waved under his fat olfactory organ.

Why had I promised Mom I'd do this? It was stupid. But I tried again. "Hey. Champ. Got a minute?"

Still no reaction. Not even a twitch of those thick, smooth brows.

I knew he wasn't deaf. Maybe a visual would help. I walked around until I caught up, paced him, and pulled the hood of my cloak away from my face. "Yo. Over here, big guy."

Jakol's eyes never left the shorn yiborra grass under his footgear, but one of his five-fingered hands made a moderately rude gesture. Mom used to do the same thing when I was a kid and bugged her too much about going topside.

Reality check time. "Looks like no one wants to spar with you, pal."

It was pretty pathetic, him hanging around, waiting for one of the Torin warriors to step over the ryata. None of them evidently wanted to. Except me, naturally, but I had to avoid that.

Or not. "Did you hear me? You're wasting your time."

A low, rough growl left his throat. It sounded something like, "Leave me alone, Terran."

I'd figured on tangling with *someone* before I left Joren. After three weeks of being politely but repeatedly slammed by the HouseClans, it might as well be the biggest, meanest guy on the planet. I shrugged off my jacket, eased my feet out of my boots, and vaulted over the ryata, landing just in back of him.

That got his attention.

He swung around and took my first punch on the chin. Solid white eyes rounded as he stumbled back against the quad ropes. Which, if we'd been actively sparring, would have cost him the match.

"Hi there." I waved. "Can we talk?"

He still hadn't quite recovered. "Mother of All Houses. Are you deranged?"

"Maybe." I spat on the grass stubble between us in time-honored Terran fashion. "Come and find out."

He glared at me and I got another curt gesture. *Go away.*

Mr. Gregarious thought he was going to dismiss me, just like my grandfather and nearly everyone else on this stupid planet had done. I could see I needed a friendly way to get things going. Jumping in the ring hadn't worked. Neither had insults, the love tap, or spitting.

So I balled up my fist, pretended he was Uzlac, and punched

him again, this time in the belly. He didn't bend over, didn't whoosh, didn't even blink.

"You're in nice shape." I restrained the urge to rub my throbbing knuckles. "You would have made a hell of a front man for me back on Terra."

"You struck me twice," he said, his gaze going from his abdomen to my face.

"And you can count, too." I darted to one side, gauged the distance between us, then kicked him in the back of the thigh.

That hurt my foot, not him.

It was risky doing this without the standard presparring agreement. If there were any Varena hanging around, I might soon be wearing my intestines for garters. Jorenians did *not* like it when you messed with their family members. Even the unwanted ones. As I had learned over the last couple of weeks.

"I shield you, Terran." He turned his back on me. "Leave the quad."

I let my temper go. "What are you, afraid to hit a girl?"

"No." He swiveled, gave me a singularly intense look; then at last he started moving toward me. "I have hit many . . . girls."

Oops. I'd forgotten there was no gender discrimination in the warrior's quad on this world.

"That's more like it." I circled backward in front of him. "So, as I was saying, I need to talk to you."

"You choose an odd manner of conversation." He lunged, and I barely got out of the way. "You are very quick."

"Yeah, I am." My knee twinged, and I wondered if I was pushing it. I hadn't had any problems since 'Gill left the Terran trader, but his warnings still echoed in my head. "Try and catch me, big fella."

We danced around the confines of the ryata like that for a few minutes. I caught him twice, solid punches, once to the throat, another against the ear. He got hold of my tunic once, but hesitated, and I twisted out of his grip. The ripping sound aggravated me even more. I'd just bought the damn thing yesterday, and it wasn't like I was rolling in credits.

Finally my knee started to ache, and I decided it was time to end it. In good runback fashion I doubled over, rolled off my palms, and came up in time to land both feet in the center of his chest.

Unused to shockball tactics, Jakol lost his balance and went down on his knees.

This was my chance. With a sphere-down shout, I jumped on his back, wrapped my legs around his abdomen, and used the momentum to force him down on his side.

Now what did I have to do to finish it? Oh, right.

I shoved him over on his back, and smacked the center of his chest with my palm. "You've lost."

A couple of Torin watching from a polite distance chuckled.

Those narrow white eyes stared up at me for a long moment. "So I have, Terran."

Jakol's claws emerged as he thrust me up and off him. They were a sure sign he was even more annoyed with me than he'd been before. Yet he didn't leave a scratch on me. I watched as he rose to his feet.

I held out a hand. "Now can we shake hands and talk?"

He gave me another irate look, made a short bow, and went back to soundlessly pacing the length of the quad.

I stayed where I was. "Jakol. Do I look like I *normally* go around challenging guys as big as you? Do the math."

He stopped, thought about it, then swiveled around and took my hand. "Why must I shake your hand?"

"It's a Terran thing." Since his claws had retracted, I shook his. His palm felt warm and callused. Manual labor, I guessed. They'd stuck him out in a field somewhere, probably thinking that would keep him tame. The morons. "See? No harm done."

"That is a matter of opinion." He released my fingers.

I straightened my garments. "Let's take a walk, shall we?"

He wasn't sold on the idea. I could tell from the way he stood there scowling, big arms folded over his chest. "Why must I walk with you, Terran?"

Because I have to destroy your illusions. Then I can leave this planet and never have to see another blue face again. Except in my dreams.

"I'm not Terran. Not all the way." I stripped off one of my gloves and flashed some claw. It wasn't hard; I was pretty annoyed myself. "Mom was a local girl."

Regardless of my shocking little revelation, Jakol took his sweet time answering. "Very well."

We jumped over the ryata barrier and pulled on our outerwear. Jakol watched me the entire time, and when I was done he said, "I would know your House."

Here we go, I thought as I pulled on my gloves, then shrugged my pack onto my shoulders. "No House. My mother—"

"Your *Clan*Mother."

They made such a big deal about what you called them. "My *mother* was kicked off this world before I was born."

He started to say something about that, then peered at my face. "What is wrong with your eyes?"

I took off my shades.

They weren't all that different from the Jorenian all-white version. Except mine were solid green—no discernible pupil, no iris, no white cornea—just green.

Jakol simply stared for a while. Then he tried to be polite. "You must be greatly admired among your HouseClan."

"Like I said, pal"—I slid my shades back on—"no House."

"Do not be pathless." He sounded like he was repeating some adult's favorite reprimand. "You said your ClanMother was Jorenian. All Joren belongs to its Houses."

"HouseClan Raska repudiated my mother twenty-five revolutions past, when she was pregnant with me. I was born outside bond. *Now* do you understand?" We were starting to attract some attention from a group of kids drawn by the activity. "Look, let's get out of here." He didn't budge. "I want you to meet the others."

"What say you? What others?"

"The others like us."

"There is no one like *us*." Jakol said it with complete certainty. So he'd already checked the census database, like I had, and discovered he was the only living half-Terran. I wasn't on it because I'd been born during Mom's jaunt to Terra.

"You can see for yourself," I said. "They're waiting to meet you. Unless you're afraid to find out you're wrong?"

A Jorenian warrior didn't retreat from any kind of challenge. Since Jakol wanted to be one, he adhered to their dumb rules.

"Very well." He turned and eyed me again. "I would have your birth name, ClanDaughter Raska."

Skalea had made it very clear I was no daughter of his House, but I stowed the explanations. "Sajora. Everyone calls me Jory." I awkwardly made the gesture adults of different Houses meeting each other for the first time used. *Should have spent more time studying this hand business,* I thought, then repeated the ritual greeting Mom had taught me. "I intend no harm to your kin, warrior."

It was the most respectful form of greeting I could have offered him. I caught a momentary flash of pleasure in his white eyes before he made the corresponding gesture. "No harm to my kin is anticipated, lady."

Poor Jakol. I was no lady. And harm would be the least of his worries after my little chat.

CHAPTER SIX

"SEEK THE DIRECTION OR ANOTHER WILL TAKE YOU."

—TAREK VARENA, CLANJOREN

As I led Jakol Varena away from the quadrilateral, I noticed how he kept scanning the crowd. I didn't know how anyone ever recognized anyone on this world—everyone was big, and had black hair and blue skin.

Then I thought of the five others waiting for us. *Well,* almost *everyone.*

I followed his gaze. Like HouseClan Raska, the Torin all stuck together, moms and dads and kids, doing the happy-family thing. They actually liked hanging out together, too; you could tell from their faces.

Then the light dawned—*he's looking for his ClanMother*—and I'd wager good credits he'd been doing that since he learned to walk. *She must be a total bitch to ignore him like this.*

Apparently he didn't spot her, because he kept looking as we passed the food pavilions. I caught the smell of something delicious, and my stomach rumbled.

Jakol must have heard, because he stopped and purchased two sticks of braised vegetables, and handed one to me. "Eat."

"Thanks." I nibbled on it, found it tasted even better than it smelled—like caramelized onion and pumpernickel—then polished it off. "That was good. What do you call it?"

"*Anh' larral.* The heart of the *anhraea* flower."

I tossed the stick in a disposal bin we passed. "You guys eat a lot of flowers."

He surprised me by asking a question. "You traveled here from Terra to find other crossbreeds like yourself?"

I didn't feel like explaining the intricacies of homeworld deportation laws. "Yeah. I got here at the beginning of *Namadar*."

He checked out a cluster of blue-skinned women discussing the merits of some silky fabric being sold in big, colorful bolts. "Is this your first journey to the homeworld, then?"

"Your homeworld, pal, not mine. I was born in *space*."

He gave me another skeptical glance. "You cannot claim Terra. Its people do not allow alien races to inhabit their world."

"We got around that." I could see he didn't comprehend. "My mother and I lived in the alien underground, in tunnels beneath the surface. We stayed there until I got old enough to train and qualify as a runback."

"Runback?" He even said it in Terran, but his palate made it sound like *Hr-roonh-vbehk?*

"I played professional shockball. First-string sphere runner, NuYork StarDrivers." From his blank look, it was apparent that he hadn't heard of me, and I felt amused by my own ego in assuming he had. "Do you know what shockball is?" He shook his head, still perplexed. "It's a sport I played. I was a professional athlete."

"They compensate athletes on your homeworld?"

Not nearly enough. "Uh-huh. Haven't you ever wanted to find out anything about Terra?"

The curious expression faded. "No."

"Too bad. Would make you appreciate Flower Central here a lot more." By then we'd reached the entrance to the subsurface labyrinth, and I indicated for him to proceed.

"We are to go in there?" He studied the mouth of the big cavern.

"Yep." I thought of all the miserable years I'd spent in the tunnels under New Angeles, damp and cold and wet, but this would be the last time. "In there."

The sudden change from light to darkness forced my companion to pause. He blinked a few times. I removed my shades and slipped them into my journey pack.

"Where are you taking me, Sajora?"

"Jory. You need to meet the others." God, he was suspicious. Of course, they'd made him that way. "There are five more."

"Five more of what? Crossbreeds?" He scoffed. "There are more than five half-Jorenian beings on our world."

My step never faltered. "There are only five like us, Jakol. We're special."

"What say you?" He caught my wrist and tugged me to a halt. His long black hair reacted to the static generated by the contact, and curled around my upper arm. "Explain to me 'like us' and 'special.' "

As moments went, it was a fairly intense one. Before, in the quad, I'd been intent on winning, and it hadn't mattered how he'd touched me. Here I was far too vulnerable. There was also something else tugging me toward him, something I hadn't felt since Rijor had died.

Not the time or place, Jory. "I need visual aids to do that, Jakol, and they're right down that way. Come on."

"Do not address me thus." He released me. "I am Kol."

Christ, they were picky. About *everything*. "You changed your name?"

He made a bitter gesture. "No one addresses me as Jakol unless they intend harm."

It must have meant something nasty in the ancient Jorenian language. "Your pardon, J—Kol." I couldn't start feeling sorry for him now. I'd never get through the next hour. "It's not far now. Follow me."

We walked into the depths of the caverns, past the interior waterfalls, and around the stalagamine pools. Funny how one underground was a lot like another, even on different worlds. The air tasted funny, the stone felt clammy and damp, and the quiet crawled over my skin. I hated caves and tunnels and small, dark places. I never felt like I could take a deep breath. It bothered me so much when I was a kid that I repeatedly swore to Mom I'd run away one day and never live below surface again.

There are many such places, Mom's ghost chided me. *Some you build yourself.*

Once we got past the tourist stuff, the sensor units dwindled until I was forced to take an optic emitter from my pack and use it to illuminate our path. It took another few minutes to pick up the markers I'd left.

"There." I pointed to the narrow entrance of a small shaft just ahead. "That's where they're waiting."

"Why have you gathered these 'others' here?" Kol looked around. "Why not meet in the open?"

"We don't want to be monitored, and what I have to say is confidential." When he stopped, I touched his arm. "They're friends, I promise."

"I have no friends, lady." He stepped away from me, and made a gesture of extreme courtesy. "Proceed as you will."

Everyone was still waiting inside.

I made a gesture of greeting to the other two females and three males. Someone had scrounged up a couple of additional emitters and placed them in the natural rock shelves, and almost everyone had taken off their hooded cloaks. In the dim light I watched my happy little group collectively check out Kol. No one showed any surprise at the color of his skin or his odd-fingered hands.

Kol was pretty hard-pressed not to return the favor.

He was right: Crossbreeds were unusual but not entirely unknown on Joren. Generally when a Jorenian bonded with an off-worlder, the resulting progeny were born with physical attributes from both of their ClanParents' species.

Like me, however, these five took that blending business to the extreme.

One of the boys was huge, more like a Hsktskt than a Jorenian. His skin wasn't blue, but a green so dark it was almost black, and possessed a pattern of subtle, symmetrical ridges you couldn't see unless the light shone over him at just the right angle. He'd either shaved his head or had no hair to begin with. I hadn't asked. His white-within-white Jorenian eyes gleamed as he sketched a ceremonial bow.

"I am Nalek, of HouseClan Zamlon," he said.

After being run out of HouseClan Raska, I'd returned to Lno and tracked Nalek's name through the hotel computer. It had taken a day to drive from Lno over to Talot Province, and another two to catch him at the HouseClan dockyards, where the Zamlon had him working like a slave. He'd been the easiest to convince.

Kol made the appropriate gesture of response, and studied Nalek's impressive musculature as though sizing him up for a ClanSpar match.

I wasn't worried about that. Nalek was a big, quiet lug who wouldn't have hurt a fly.

The smallest person in the group stood at Nalek's side, a little in his shadow. The fragile-looking female had huge, iridescent offworlder's eyes and skin paler than mine or Kol's. As she peeped out to make her timid welcome gesture, two stunted winglets appeared on either side of her shoulders. You could see the beginnings of feathers sprouting all along the shorn appendages before she jerked them back down and folded them tightly against her spine.

"Welcome." The end of her sleek black braid touched the floor of the cave as she inclined her narrow head. "I am Galena, ClanDaughter of HouseClan Nerea."

Her voice still sounded shaky. Since I'd smuggled her out of Pnoek Province, she'd been nervous as hell. Couldn't blame her, seeing as her overprotective family had been ready to declare me ClanKill about an hour after they'd met me. I got the feeling their hostility wasn't because of my Terran hide, but to keep Galena secluded, maybe to safeguard her from everyone who might make fun of her—or them. I'd been forced to smuggle her out of the Nerea pavilion in the middle of the night, and they began pursuing us before we left Pnoek Province the next day. Far as I knew, her kin were still searching for us.

"You have wings," Kol said.

Before Galena could reply, a mocking tenor rang out. "ClanSon Varena has the gift of observation."

Kol swiveled to stare at the square-bodied, compact form of the male who had spoken. Half again as wide as he was tall, he wore the briefest of garments, mostly because of the thick scales of hard blue exocartilage covering his body. His face was paved with smaller, thinner versions of the same. He had two Jorenian eyes, but they were lidless, and he sported an extra pair of limbs, complete with six-fingered Jorenian hands, between his shoulders and hips.

"Osrea," the reptilian crossbreed said, but he skipped the welcoming gesture. "HouseClan Levka."

Frankly, Osrea had been a complete jerk about joining the party. It had taken a week, but I'd finally found him living in what amounted to a hole in the ground just outside Moalan Province. However isolated and dismal, it had still been *Osrea's* hole in the ground, and he hadn't wanted to abandon it. It had taken outright bribery to persuade him to come with me.

The last two stood at the very back of the cave. One was a female with purplish skin and an unruly mass of thick yellow hair. She gestured first to the third male, who still wore his hood. "This is Renor, of HouseClan Xado. I am Danea, of HouseClan Koralko."

Renor and Danea, surprisingly, had found *me*. They'd met me one morning coming out of the Marine Province tourist lodge, where I'd been staying with Galena. At first I'd thought they might be Nerea kin, until Danea produced a datapad with a copy of the inquiries I'd sent to both their HouseClans, asking for a meeting with them. She claimed they'd tracked me by contacting the tourist lodges and asking if they had any Terran guests. I was satisfied to leave it at that, especially after I'd accidentally brushed against Danea, then had gotten a good look at Renor. They both spooked me. A lot.

"Kol, of HouseClan Varena," my companion said, and stiffly bowed to the group.

"Have you no horns?" Osrea asked, and his serpentine tongue lashed out to taste the air. "No auxiliary limbs, no exotic appendages?"

"Why? Do your eyes not function?" Kol shot back.

He and Snake Boy both had chips on their shoulders the size of star shuttles. Teaching them to play nice was going to be a job.

"Okay, Os, knock it off." I'd been polite with everyone, but it never worked with him, so I matched his blunt aggressiveness instead. "You can see he's just like me—half Terran."

He snorted through his recessed nostrils. "You have my condolences."

"Kol." I touched one rigid arm. "Osrea isn't trying to start a fight." Of course he was, but I couldn't let him do it. This was supposed to be a solemn occasion, not a brawl. I turned my head to address the others. "Come on, everyone; let's cease fire, okay? Kol is the last."

"The last of what, lady?" Nalek's rich, deep voice echoed like a mellow bell in the small alcove. "You have yet to fully explain the purpose of this gathering to any of us."

"Sit down," I said. "My formal Jorenian is lousy; this is going to take a while."

Kol reluctantly joined the circle as the others moved in and sat on the cold stone floor. I stayed standing until they were settled, then took one more mental head count.

Jakol. Nalek. Galena. Osrea. Danea. Renor. And you, Sajora. Wherever souls went after death, my mother's was definitely dancing. *At last, the seven complete. You must tell them all as one.*

"Well?" Osrea's tongue flickered with impatience.

Because of the promise I'd made, I had to say this in their language, in front of all of them. Then I was done—done with Joren and Jorenians forever. I took off my wristcom and tossed it aside.

Here we go, Mom. Down at the line of scrimmage.

"I am the birth child of Kalea Raska." Even with a couple weeks' practice, it was still hard to wrap my tongue around their liquid phonetics. Then I had to deal with doing the hand stuff with every other word. "Know you my moth— my ClanMother?"

Everyone nodded along with Kol. Mom had been a well-known exobiologist, honored as ClanJoren, or a daughter of all Houses. At least until she'd been thrown off Joren.

What did they call death? "My ClanMother has embraced the stars."

Everyone smiled, and Danea said, "We wish you joy, Clan-Daughter Raska."

"I thank you for that." No, I didn't. I never understood this party-when-someone-dies attitude the Jorenians had. "My mother's diversion was unexpected. I come before you as her Speaker." That was roughly the equivalent of verbally presenting Mom's last will and testament.

Everyone stopped smiling.

"You must not Speak to us, Sajora." Kol got up and held out a hand, as if he intended to clap it over my mouth. "Speaking must be before the HouseClan. You must return to the Raska to perform this ceremony."

"No, it's not like that. Sit and attend to me." I motioned him back down. "The Raska are not my HouseClan. You six are."

There were different reactions to this statement. Osrea erupted into laughter. Nalek looked appalled. Galena covered her face with her hands. Danea's hair fanned out. Renor made no movement or sound.

Kol simply stared. I knew what he was thinking. There was no possibility we could be ClanSiblings, unless—

"Who says that Terrans cannot be amusing?" Osrea clutched his side and laughed again. "A good joke, that one. Tell another, Sajora."

Smart-ass. "Okay. What is the name of your biological sire, Osrea?"

That shut him up.

"Well, what say you?" When Osrea refused to respond, I turned to Nalek. "And you, Nalek Zamlon? What is the name of the male who sired you?"

"I do not know." He ran a huge hand over his dark, bald head. "I was never told."

"Sajora—Jory." Galena's pitiful wings spread wide, as if she were preparing to take flight. She just might get a chance, now that she could grow her feathers back in. "I beg you stop this and return to the Raska. No good will come of this . . . Speaking."

"You think not, ClanSister?" I crouched down and took her

small hands in mine. She was so damn brittle and sensitive; I was almost tempted to let them all hang on to their illusions. But then I'd be just like everyone else on this stupid world. And I had promised my mother. "Who was your sire?"

Pain etched her delicate features. "I have no knowledge of him, except that he was an avatar."

"My sire is Terran and dead." Kol sounded like he'd heard enough. "Finish what you must say, Sajora, or I leave now."

He meant it, too. I sighed. "Okay." I got up and made what I hoped was the correct gesture for expressing the dying wish of a Jorenian. Mom hadn't exactly been thrilled about teaching me that. "I Speak for the daughter of my HouseClan, Kalea Raska. Her words were given to me, to be brought to the other six ClanChildren of Honor. I bring them with joy."

I could feel my mother smiling inside my head. *Bring them with joy, with my joy.*

"The *ClanChildren* of Honor?" Danea pushed herself to her feet and took an aggressive stance. Her light hair ballooned around her dark face. "You babble nonsense. There is no such House."

"Please, Speak on," Nalek said, now looking very intent.

I closed my eyes for a moment, then Spoke for my mother. " 'I vowed to bring the truth to you six, but I am in exile, and it is not my path to return to Joren. Thus I have given Jory the task and made her vow to Speak in my place.

" 'Twenty-six revolutions past, there was an attack on a Jorenian vessel en route to the homeworld. A raider fleet captured the ship. Many paths were diverted. Seven Jorenian females were sold to slavers on different worlds. After some time, our HouseClans located and liberated the hostages." I paused. This was the part they *really* weren't going to like. "And our children by our slave owners.' "

Kol shook his head slowly. "No. No."

I was right: No one had ever told them the truth about their fathers—not even in confidence.

I pushed on. " 'The Ruling Council could not hold the children responsible for the dishonor to their ClanMothers. Each HouseClan agreed to accept the seven, and vowed to regard them as if born from bond.' " *Which hadn't happened, from what I'd seen. So much for keeping those solemn vows.*

Osrea's tongue flickered rapidly. "Impossible. My ClanMother would not endure life as a slave."

" 'As was tradition before the path of Tarek Varena, the seven of us were Chosen upon our return to the homeworld. Six children were brought to their native HouseClans. I, Kalea, was in the last stages of pregnancy upon my rescue. I refused to honor the Choice made for me.' " *Thank God.* " 'My sentence was repudiation and banishment. The records of our ordeal were destroyed, and each HouseClan swore an oath never to seek the honor of ClanKill against the slavers, or even to speak of what had occurred among us.' "

"Our ClanFathers were not slavers." Danea made a low, growling sound in her throat, and her claws emerged. The air around her started to glow with a weird yellow light. "You lie."

" 'I charge my child with my last request: Seek out the other six ClanChildren of Honor. Make known to them the dishonor inflicted upon their ClanMothers, so that they may choose their path. Farewell and safe journey. I embrace the stars.' "

Nalek pushed me out of the way as Danea sprang. She came up short, as though surprised by what he'd done. Then she started after me again.

"I shield this one," Nal insisted, but she tried to go around him. He ended up wrestling the yellow-haired girl to the floor of the cave. "Cease this assault, ClanSister!"

The interesting thing was what was *not* happening. The one and only time I'd accidentally touched Danea, I'd gotten a mild jolt, the same kind I'd have received as a penalty for an illegal block during a game. She hadn't gone into detail after Renor had helped me off the ground, but I gathered her dad's species produced a natural bioelectrical charge. After that meeting, I started wearing my insulating thermals again. Only now Nalek had her pinned on her back, and wasn't getting any sort of a charge at all.

"Do not claim me kin, Houseless son of a Hsktskt!" Danea yelled. "Release your shield so I may gut this deceiver!"

I didn't need the big guy protecting me. "Let her go, Nalek. I can take her." I hoped. My thermals would absorb a fairly serious jolt, but not more than two or three of them in a row.

"I will cook you where you stand!" Danea pushed Nalek off with a nice move and got back up on her feet.

"Yeah, but the truth hurts more." I assumed a defensive stance. "Come on. Come and get me, Sparky."

"Danea." Renor finally stepped forward and removed his hood. The sight of his glittering face made everyone gasp. "No path will be diverted on this day."

Over Nalek's shoulder, I saw Kol. He couldn't move his eyes from the crystalline facets of Renor's face. I couldn't blame him. It was as if Ren's entire head had been encased in plas, then hewn to vaguely resemble a Jorenian. His eyes were not white or offworlder, but thin, horizontal black slits.

"She lies!" In spite of her tone, Danea edged back against one wall. Was she afraid of Renor?

Nalek helped me to my feet, and I dusted myself off. "I do not lie, ClanSister." Tired of sputtering through their lingo, I groped for my wristcom, slipped it on, and activated it. "Look, we were all sired outside bond, by men who were slavers. Or rapists. Take your pick."

"You said we had to come together, that what you would tell us would change our lives." Renor's voice matched his outside—hard and cold. "This was what you meant by that?"

"Your own families never cared enough to go after the slavers who did this, or even tell you the truth so you could," I said. "And what *did* they tell you? That your fathers were alien tourists or visiting scientists? How did they explain the fact that they weren't around anymore?" I watched their faces, saw their eyes shift.

Kol rubbed a hand over his face. "Perhaps they meant to protect us."

I snorted. "Right. More likely they wanted to hide you, like a really ugly dog no one has the heart to put down."

"What is a dog?" Os wanted to know.

"A domesticated animal on Terra. People keep them as pets. Most dogs only need regular feeding and a pat on the head now and then to stay happy." I met Kol's gaze. "The unhappy ones get chained in the yard."

"We may not be the most honored of our Houses," Nalek admitted, "but our kin swore to protect us. That does mean something, Jory."

"Well, my mother didn't take their stupid oath, and she thought it was really important that you know the truth. Now you know." I spread my hands. "That's it. I'm done, and I'm leaving now."

"You have another agenda. You are Terran." Danea turned her head and glared at Renor, then moved forward away from the wall. "Your kind are hostile bigots who care nothing for Joren or us."

"What is it with you people?" I yelled. "I come here to do the decent thing and keep a promise to my dead mother and you think I'm out to swindle you!"

More yellow hair spiked. "Are you not? Your people hold offworlders in utter contempt."

"And your people are so wonderful?" I swept a hand around the room. "Nalek, did anyone ever offer to let you build a ship instead of killing yourself hauling the materials to and from the dock?"

His heavy jaw sagged, then locked. "No one knows."

"I knew from the moment I saw all those vessel designs and drawings you put up in your quarters. They were gorgeous." I rolled my eyes. "So why aren't you building them? Are the Zamlon all blind?"

Nalek glanced down at himself. "My physical form is more suitable to manual labor than craft design."

"What about your brain? Why doesn't that count?" I went to Galena. "Birdie, who's been plucking your feathers all these years?"

Galena studied her footgear. "My ClanMother wanted my appearance to be similar to that of my ClanSiblings."

"I'm surprised she didn't have the wings amputated while she was at it. And Os—before you made that dugout in the ground, where did your family have you living?" He started to say something, then hissed. "It's okay; your ClanCousin told me. In a shack, as far away from the HouseClan pavilion as they could stick you, practically in the middle of their *t'lerue* herd. Keeping you there made the buyers from the other provinces a little less nervous, he said."

"They think I am part Hsktskt." Os's sneer became a painful grimace. "I am not, you know."

"I know. They know—but they didn't bother to tell anyone else that, did they? They just stuck you out there with the cows." I turned to Danea, whose face had darkened to a beautiful shade of royal violet. "Sparky, what did they do to you and Renor?"

She bared her pointed teeth. "Call me that name again and you die."

"Okay, we'll skip you." I turned to Kol. "But not you. How many times have you applied to join planetary militia? How many times have they rejected you and taken someone else from your HouseClan? Someone younger? Someone you could stomp into smithereens in the warrior's quad?"

"Five times," he said, staring back at me through hot white eyes. "I stopped applying after the fifth dismissal. How did you know?"

"One of the Torin mentioned it, along with what a shame it was that you weren't allowed to serve the Varena. Too bad they didn't back up their regret with action." I gazed at the other resentful faces. "The only thing you're guilty of is being born, and yet you all act like you deserve this kind of treatment. No questions, no confrontations, you simply shut up and take it." I rubbed my temples. "Hey, that's your choice. I've kept my promise to my mother, and now I'm out of here."

Galena's thin hand briefly touched my arm. "Where will you go, Sajora?"

"Oh, I'm going to do what the HouseClans should have done twenty-five years ago. I'm going to find the raider who kidnapped and sold our mothers, and I'm going to kill him."

"Declare ClanKill," Nalek corrected me.

"Whatever. You guys can do whatever you want, but don't kid yourselves—these people hate you. They will always hate you."

They all stared at each other, then me. Even after everything that I'd just told them, they acted as if that were another stunner.

"Surely we are not hated. Surely . . . not that." Galena put a slim hand to her throat. "This dishonor was none of our doing."

"One look at you, Birdie, and they remember how you were conceived. You know what this human asshole told me before I was deported from Terra? 'Alien blood always shows through.' " I took a deep breath. "They see you as the sons and daughters of slave owners. They can't help it, not with the way you all resemble your sires. Oh, they'll protect you, and feed you, and if you're lucky, they won't chain you in the yard. But to your kin you are walking, talking reminders of a crime that went unpunished."

"Perhaps you are right, Terran, in what you do." Danea paced a short distance around me. "If we go and avenge the dishonor done to our ClanMothers, then our people, our HouseClans will have to accept us."

Now it was my turn to gape for a moment. "Ah, no. This is *my* business. I don't need any help. I don't want any help. Especially from you."

"Obviously you do not." She turned to the others. "Yet it is apparent that we are in need of yours."

CHAPTER SEVEN

"Lead not for glory, but so that others may find your path."

—Tarek Varena, ClanJoren

Namadar had been over for nearly a week when Uzlac finally got clearance to leave Joren. By that time I'd handed over to him all the credits I had left. I'd also worn out my welcome in Marine Province. HouseClan security officers had questioned me twice, and both times I'd pretended not to know where Galena or the others were.

Security didn't pretend to believe me the second time.

"Galena's ClanMother has shielded you against the rest of the Nerea," one of them told me. "All she wishes is her ClanDaughter's safe return."

I studied a spot of mud on my boot. "I'll pass the message along if I see her."

The officer frowned. "You risk much by interfering in matters of privacy, Terran. We can always detain you for further questioning."

"True." Time to bluff. "And if you do, I've instructed my legal counsel to release a full statement regarding the *MoonWave* incident and everything I know about the ClanChildren of Honor." I *tsk*ed. "Some very volatile stuff." Actually, nobody would care—except the Jorenians.

"Such statements can be suppressed."

"On Joren, sure. Everywhere else in the quadrant? Not a chance. Still"—I lifted a shoulder—"if you guys don't care about that kind of exposure, toss me in a cell. Who's really going to care about an old scandal like this, anyway?"

They released me and a few hours later the Ruling Council sent me a politely worded request that basically meant, *Get off our planet.*

Which was why Uzlac got permission to depart, of course.

My first and only stop on the way to Uzlac's star shuttle was Kol's place. His ClanMother, Qelta, greeted me at the door panel with the customary disapproving glare.

"Hi." I leaned an arm against the interior panel frame so she wouldn't key the door to shut. I hoped. "Can Kol come out and play?"

Her mouth crimped. "He is resting."

"I'll wake him up." I pushed past her and headed back to his chamber.

Kol was folding his garments and stacking them next to a couple of open cases, I saw as I stepped in through his door panel. "You ready to do this?"

"No." He went on packing.

None of them were; I knew that. The whole group had argued about it for hours before leaving the caves. I hadn't been thrilled by the idea of having six partners. And they hadn't taken much to my idea of traveling to Reytalon for training first.

"The seven of us, become blade dancers?" Osrea laughed so hard he ended up on the stone floor, clutching himself with all four arms.

"It is not possible." Nalek didn't laugh. "I am not sure I could kill anyone, even with such an education."

Danea could, and I had the feeling she put my name on the top of her after-graduation list. She glowered at me. "They will not accept us. We have no funds, no references."

I told her what the dancer had told me.

Kol didn't like that. "I will not be a mercenary for hire."

"It's an option, Kol. Not an obligation."

Galena reached over her shoulder and plucked at the top of one winglet. "Will they expect me to fly?" I pointed out that she had no feathers. "If they grow back, will they expect me to fly then?"

Even Renor, who hardly said a word about anything else, had to be a pain about it. "What if we kill another student?"

I threw up my hands. "It's a place that *teaches* killing, Ren. I'm pretty sure they're prepared for that."

My thoughts snapped back to the present when Qelta appeared in the open doorway of Kol's room like a nosy chaperon. Her suspicious gaze zeroed in on his case. "You plan to take a journey, Kol?" The words and gesture were polite, but her tone said, *Not with her, you're not.*

"Yes, ClanMother." He removed one last stack of garments from his storage unit and handed them to me. Knowing our time was short, I dumped them on his sleeping platform and started shoving them into a case. "My thanks, but I will do it." He folded each garment carefully before placing it in his open case.

Brother, was he finicky.

His ClanMother's voice became forcefully cheerful. "Where say you journey, my ClanSon? Piran Province, to celebrate your ClanCousin Tborna's bond? Your ClanFather and I would travel with you—"

"We leave Joren."

Stunned, Qelta gasped and pressed a hand to her breast. At the same time, if looks were lethal, I'd have been on the floor bleeding from every opening.

She found her voice. "You must not. You *cannot.* You are too young."

Kol secured the case latches. "I am of majority age."

She wasn't giving up. Not without a fight. "You have another revolution of education to complete, my ClanSon, before you can assume even a junior security officer's position."

"The Varena have made it clear I am not to expect a position on any of their vessels. No other House will hire me. The militia have rejected every one of my applications." He stared at my shades and the side of his mouth curled in a bitter half smile. "I will complete my education elsewhere."

I silently willed him not to say anything more on that subject. If his ClanMother learned where he intended to finish his schooling, I *would* be dead.

The hand over Qelta's heart curled into a fist. "This is her doing, is it not? This Terran. Has she lured you into taking this journey? Promised you rewards and glory?"

Kol said, "No."

At the same time, I said, "Yep."

"I will not permit you to leave us, Kol." His mother crossed the distance between them and grabbed his big arm. "Do you hear me? I forbid it."

"You cannot." Kol stared steadily at her. "It is my right." He shook off her hand. "Who was my sire?"

I pressed a hand to my forehead. *Oh, Christ, Kol, don't do this now.*

"Your ClanFather is Nla, of course."

"No." He made a vicious gesture. "I was born outside bond. I have no ClanFather."

"Kol." Qelta's eyes bulged, and she reached out to him, flinching when he jerked away. "Never say such a thing. Nla has been your ClanFather since we Chose."

That was the whole problem. Qelta and her bondmate had yanked Kol out of HouseClan Varena, exiled themselves among the Torin, and never explained that they'd done it because he was the illegitimate son of a Terran slave owner. They'd just expected him to swallow whatever waste they handed him about it when he asked questions. Like now.

It pissed me off.

"He knows about the *MoonWave*, Qelta." I stepped between them and faced her. "I told him about his sire."

Her whole body tensed, and her claws emerged. "I should never have allowed you to pass over our threshold."

She didn't scare me. "Too late."

"Who was it?" Now Kol came around me and advanced on Qelta. "Who violated you?"

She gave me a quick look, and her skin turned a chalky color. "I know not of whom you speak."

"He was Terran; that much I know." He swung a hand toward me. "Was he Sajora's sire, as well? Tell us!"

That was something I hadn't considered—could Kol be my half brother? Afraid Qelta might tell him more than I wanted him to know, I grabbed his arm. "Let's just get out of here, okay?"

"Yes. You should leave."

The three of us looked around to see Nla standing in the doorway. The man who had adopted Kol was a tall, dour man who had said exactly three words to me over the last week. He never went near his bondmate, and Qelta seemed to avoid him just as fervently. From what I'd observed, Nla seemed happiest contemplating the intricacies of journey philosophy, or working in the vast Torin botanical fields.

He also treated Kol like he had a permanent bad smell.

"We are leaving, Nla."

I guessed that was the first time Kol had ever addressed his ClanMother's bondmate as anything other than ClanFather. Qelta gasped as though struck, then began to weep.

"Kol." Nla's claws were out, and he appeared ready to gut his own adopted son. "You do not honor the House." He gave me the usual disgusted look.

Qelta shuddered. "This female has polluted his mind with her outlandish tales."

"The outlandish truth," I said. *Well, mostly.*

"I am leaving." Kol jerked his cases from his sleeping platform, handed one to me, then walked around Qelta.

She clutched his arm at once. "Do not take this path, my ClanSon. I beg you."

Kol paused. "Was it truly as Jory told me? Was I sired by a slave owner? Sired in rape?" Qelta closed her eyes, swallowed, and gave a jerk of her head that passed as a nod. "I would have his name from you, ClanMother."

"No." Her white eyes snapped open, wide and furious. "I will not speak it."

My ClanBrother's face turned to stone. "I will find him anyway."

Qelta looked from Kol to her bondmate, then slowly let her hand fall to her side. She covered her face and sobbed.

"Good-bye, ClanMother." Nla blocked his path, and Kol met his gaze. "I would leave now, my ClanMother's bondmate."

A huge fist seized the front of Kol's tunic, and the quiet

man's face turned a little purple. "You have never brought honor to our name, Jakol."

"Indeed." He didn't blink. "That will soon change."

Nla released him and pushed past us to get to his bondmate. He enfolded her shaking form in his long arms, more to hold her back than to show affection, then glanced at Kol. "I would wish you safe journey, ClanSon of my bondmate." He gave me the gesture that meant something like *Good riddance, trash*.

Keeping his dignity, Kol made the gesture a dying man performed for his kin. "Walk within beauty, HouseClan Varena."

Then, without another word, he stalked out of the only home he'd ever known.

I regarded the couple with a tinge of sympathy as I went to follow him. "Don't worry. I'll look out for him."

Qelta suddenly lashed out at me, and the tips of her longest claws caught me across the chest. Four parallel slashes appeared in my tunic and began to darken.

I pressed my hand against the scratches, then wiped the blood on the front of my trousers. What a *bitch*. She would have taken a second swipe, but Nla caught her arms.

"Yeah. Well." I smiled. "It's been a real charge."

We got on the local glidebus and sat all the way in the back where we wouldn't be bothered. Kol didn't notice the wounds until we were halfway to Transport. Then he did. At the top of his lungs.

"*You are hurt.*"

"It's nothing." I stared out at the incredible expanse of sweeping planes all around the glidepath. I wasn't going to miss the people, but Joren itself had grown on me. What would my life have been like if I'd grown up here? "Your mom has pretty decent reflexes, by the way."

"Let me see." He was already yanking aside my scarf, pulling open my tunic, and poking at me.

I slapped at his hand. "They're just scratches, Kol. I've gotten worse during pregame warm-ups. Relax, will you?"

"They are *not* mere scratches." That was when he noticed the scars around my neck. "Mother of—*What is this*?"

"Stop yelling; it's just a ligature mark." I was distracted by his smell for a moment. Our sweat had no odor, but there was a strange scent coming off his skin. The closest match I could think of was rain on pine needles.

"Ligature?" He lifted my curls away from my neck and found out it went all the way around. "Someone tried to strangle you?"

"Couple of times." I ran my fingertip along the shiny badge of honor I'd earned as a six-year-old. "Occasionally Terrans get a little hostile about alien kids running around their world. A group of older kids caught me topside and tried to hang me with component wire. Only I was too heavy, and it snapped."

His eyes narrowed, the tips of his claws emerged, and for a moment every sound around us seemed to disappear.

"I have worse scars from playing ball." I broke the spell by giving him a cocky grin. "But I'm not showing you *those.*"

He unfastened my tunic and pressed his sleeve over the worst of the four parallel gashes on my sternum. Other passengers were also paying close attention to that region of my body, but not because of the wounds. From the gestures a few made, it was my small, pale breasts that had them mesmerized.

I didn't know why. Jorenian women all had a lot more stock than I did in that department.

Maybe it's because my nipples are pink instead of blue-green. "Kol?" He looked up. "Mind covering up my chest before the driver crashes this vehicle into something?" His hand pulled my tunic back down over the scratches. Disappointed gazes turned away. "Thanks."

He stared out the view panel for a while. Then he spoke, startling me. "I was not going to travel with you. Yet had I chosen differently—"

"You'd still be in the same rut. I understand." And I did. It had taken deportation, an assassin, and coming here to wake me up. "Terrans call it being between a rock and a hard place."

Kol chuffed out something that resembled a laugh. "An appropriate analogy. The others, they will be waiting at the shuttle for us?"

If they weren't, I'd never convince Kol to go with me alone. These people were clannish to the extreme. "Probably."

When we arrived at Transport, I bypassed the main terminal and took Kol directly to Uzlac's vessel. As I'd expected, he was not impressed. He stopped dead in his tracks and dropped his cases on the docking pad.

"You contracted space on a *Ramothorran* star shuttle?"

"Yeah." I rubbed my chest, which stung and was still oozing a little blood here and there. "So?"

Kol picked up his cases and did an about-face. "Good-bye, Sajora." Off he went in the opposite direction.

"Kol. *Kol.* Hold on." I had to trot to catch up. "Give me a break, will you? I'm not wealthy. Under the circumstances, it was the best I could do."

He kept moving. "Do not tell me what was the worst, I beg you."

Suddenly I got mad. Very mad. I'd gone through a great deal of trouble to get to this stupid planet. Spent every credit I had. Listened to a lot of threats and nasty accusations, gotten myself chased, jolted, and clawed. Just to keep a promise to a dead woman who would never know anyway. Now Kol was walking.

Some surrogate brother *he* was.

"Fine. Stay here and be bored to death for the rest of your miserable life. Just remember, I gave you an out and *you* ran away from it." I threw my wristcom at him, and it bounced off his back. "Did you hear me, you blockhead?"

He hesitated, rattled something off in Jorenian too fast for me to follow, then stalked away.

God, he's really going to bail on me. But why was I surprised? That's what everyone did.

"Yeah? Well, same to you, pal." I grabbed my case and headed for Uzlac's docking ramp. "As if I needed this shit in the first place."

"Jory."

I ignored him. In a remote sense I was relieved. I didn't want them with me, not when I found Kieran. I didn't want anyone with me when I found Kieran.

"*Jory.*" Kol's voice thundered across the space between us. "They come."

My fury subsided as I turned to see the other five approaching us. "Tell them whatever you want. I'm leaving." I made an obscene Terran gesture that used to get me a good smack from Mom once she'd discovered what it meant.

"*I* did not arrange this," Kol said. "*You* tell them."

"They'll figure it out." He was being such a jerk. "I'm getting on this shuttle and jaunting to Reytalon. Go back to chopping grass and being ignored in the quad. Maybe in a few years they'll give you a longer chain."

I entered Uzlac's main cabin and found the agitated trader waiting for me. Not exactly an inspirational vision.

Ramothorrans were blocky humanoids with intense black eyes and single-nostril noses. Uzlac was a bit overweight, even for one of his kind, and his bulk made the deck panels groan beneath his stumplike feet.

As usual, he got right down to business. "Go now, fem Jory."

I gave him my case. Maybe, just maybe, something I'd said would budge Kol. "You can wait another minute, Uzlac. It won't kill you."

"Giant blue people will. Go now."

"All right." I went back down to the exterior entrance panel and stuck my head out. My six surrogate siblings were having an intense discussion and, from the looks on their faces, not one of bliss and harmonious agreement.

They'll probably argue about it until the yiborra grass grows up and over them.

"We're firing engines!" I shouted. "If you're coming with me, get on the goddamn ship!"

The group's hands got busy, voices rose; then Nalek, Danea, and Renor started toward the ramp. They said nothing to me as they filed past. Galena made an apologetic gesture to Kol and tagged after them. Osrea did something less tactful but also came on board.

That left Kol standing there, tall, proud, and resolute.

And alone.

" 'Bye, brother!" I yelled, and made as if to go back into the main cabin.

At last Kol stomped over. His feet nearly left dents in the ramp alloy as he boarded Uzlac's shuttle. "I do not *wish* to be your ClanBrother," he snarled as he went by me.

"Sorry, can't pick your relatives." I came up behind them and fixed Uzlac with my best evil eye. "You clear out those cabins like I told you?"

The Ramothorran made a wet, disgusting sound with his lips. He was examining poor little Galena with so much interest I thought I might have to repeat myself. "Yes. Clear. Go now."

Might as well set some ground rules right here and now. I walked over, grabbed the trader by the front of his soiled tunic, and pushed him against the closest interior hull panel. You had to be firm with the hired help, especially when they were Ramothorran lechers.

"Pay attention, you pig. These people are my *family.* You treat them with respect, or I'll dismember you and toss the pieces out a pressure lock." I tapped one of my claws against what passed as his chin. "Understand me?"

"Yes." Uzlac glanced down, saw the rest of my claws, and gulped. "Go now."

I shoved him toward the helm compartment, then looked at my "family." They were all, with the exception of Renor, whose face I couldn't see beneath the hood, gaping at me.

I sighed and waved. "Come on. I'll show you where we'll be spending the next two fun-filled weeks."

Jorenians were much more fastidious than ex-shockball runbacks, so I expected them to be dismayed by the cramped, dingy little quarters Uzlac had allocated for us.

They were.

"There are only two rooms," Nalek said, a frown creasing the ridges on his dark green face as he studied the limited furnishings and lack of deck space, all covered with the dust of neglect.

"Boys in one, girls in another," I said as I tossed my case in through the adjoining door panel. The movement made my chest burn. "Galena, give me a hand in here for a minute."

To save explanations, I made sure I closed the door panel before I stripped off my tunic.

My little ClanSister gasped when she saw the purple-encrusted gashes Qelta's claws had left across my sternum.

"Jory!" She pushed me down on one of the sleeping platforms. "Who did this to you?"

"Kol's mom." I opened the first-aid kit I'd taken from my pack and gave the mini hand-laser to her before I flopped backward. "You're going to have to seal up the bottom one; it's pretty deep."

"Seal?" Horrified, Galena dropped the instrument next to me. "You mean *burn* you?"

"I'd do it myself, but I can't see it all that well." I beckoned to the bare wall panels. "No mirrors in here, either." I rolled my eyes at her expression. "It'll only take a minute, sweetheart. Just point, press, and drag."

"I will be sick," Galena said, pressing thin fingers to her mouth.

"Fine. Go puke." I folded my arms behind my head. "I can wait."

She made an exasperated gesture and gingerly picked up the hand-laser again. "Very well. But do you not want something for pain?"

I smiled. Poor kid, she had no idea who she was bunking with. "No, thanks. Go ahead, I promise I won't cry."

It took her a few minutes to gather her nerve, but Galena did a nice job on the gash. When she was done, she rushed off to the adjoining lavatory and spent a few minutes in there before emerging, her wingblades moving in an agitated fashion.

"Everything come up okay?"

She gave me a look of mild dislike. "Nothing hurts you, does it, ClanSister?" Without waiting for a reply, she left the room.

I waited until I was sure no one else was coming in, then pulled the pillow from under my head and covered my face with it. I pressed one hand to the throbbing agony in my chest that had nothing to do with the gashes. Getting kicked off Joren had hurt more than I'd thought.

So I broke my promise to Galena, and cried.

* * *

I must have slept for half a rotation, because when I got up the room was dark, and Danea and Galena were occupying two of the other platforms. I slipped out of our room and found three of the boys were doing the same thing. How I didn't know, considering the volume of Osrea's snoring nearly shook the wall panels off their stud supports.

Everyone was accounted for, except Renor.

Since I didn't think my crystalline ClanBrother was chatting with Uzlac, I went on a silent reconnaissance mission. It took hunting through three decks before I finally located him in front of one of the data storage units, staring at the display.

Whatever he was looking at was strange. A vista of some alien world with a black sky and icy, barren mountains. It didn't look like anything could live there.

"I believe this may be the homeworld of my sire," Renor said, making me yelp. "Slaves are bought and put to work, mining beneath the surface."

"Looks cold."

"Surviving on the surface requires special envirosuits or a silicon derma, like mine." He deactivated the screen and turned around, allowing the hood to slip from his head. "You are recovered from your injuries, Jory?"

"Yeah, I'm fine." I leaned against a wall panel and tucked my hands in my tunic pockets. "What are you doing up? Can't sleep?"

"I do not sleep." Renor's features didn't lend themselves to much animation, but I thought I saw humor glinting in his cat-slit eyes. "Another of my sire's legacies. Apparently his kind don't require rest intervals."

"You must have been a pain in the ass as a baby," I said. "Seen our captain?"

"The Ramothorran is, like the others, asleep." Renor rose and gestured toward a corridor. "Walk with me?"

I walked with him. We went from the lower deck up to the helm, where Uzlac had left the ship on autodrone.

"You haven't had much to say about our plans, Ren." I gave him a sideways glance.

"True."

"You don't say much at all."

He studied the navigational array for a moment. "Conversation is difficult for me."

I leaned over his shoulder and checked out the ship's current heading. "There's no reason to be shy. You're among family now." *Sort of.*

"I am not shy; I have had no practice. Danea taught me to speak but a year ago." He made a brief hand gesture, as awkwardly as I might have. "Before that, I did not know how."

I stared at him. "What about your family? Didn't they talk to you?"

"No." He pressed a few buttons and the programmed flight coordinates appeared on the display. "The Xado kept me sequestered, Jory."

"You mean *imprisoned?*" He nodded, and the interior lights made his crystalline skin sparkle. "How long?"

"Twenty-four years, two cycles, six weeks, three days, seventeen hours, and thirty-one minutes."

I sat down fast. "Why?" Then, in a tight voice, "Because of the way you look?"

"As a young child, I was unwilling to follow their instructions. Contact with my derma causes wounds. Other . . . reasons. The Xado decided that I would cause no harm if I lived in meditative isolation from the HouseClan." His cheek glittered as it tightened. "They were correct. I caused no harm."

Meditative isolation. They'd locked him up all his life, and called it that. I wanted to go back to Joren and kill someone, *anyone,* named Xado. "How did you meet Sparky?"

The facets around his mouth crinkled into what had to be his version of a smile. "Danea found me during a visit to my HouseClan pavilion. It was she who . . . freed me from the chamber, and helped me to escape the Xado." He studied the display. "Jory, I believe these computations are in error."

I was still thinking about Renor spending his entire life in a cell, never learning how to talk because there was no one to teach him. "Huh?"

One glittering finger pointed to a series of numerals. "This will not take the ship to Reytalon. I believe Uzlac means to bypass that system and take us elsewhere."

"Elsewhere as in where, exactly?"

"A moment, I must verify the ship's position." He took another reading and studied it. "The fifth planet in this system. It is identified as Garnot."

The Garnot scandal had exploded only a few months ago, but the League had notified every member world that the alleged artists' colony was actually a slaver depot. Even I had heard about it. I sprang to my feet. "Where are the steering controls on this thing?"

Renor shook his head. "I do not know how to override these commands. We will need to obtain them from the captain."

"No override." Uzlac appeared in front of us, a wicked-looking pulse rifle in his hands. "We go Garnot."

"Oh, you'll get there, one piece at a time." I would have tackled him, but an unexpected cold grip stopped me. "Let go of me, Renor."

"He will kill you."

"Don't want to." Uzlac grinned. "But will."

Renor pushed me behind him and advanced on the Ramothorran. "I would not advise you to try, Captain Uzlac."

At the same time, Kol and Nalek showed up. I heard claws emerge, and a low, guttural sound. I don't know why, but I was pretty sure the sounds came from Kol.

Uzlac swung around and jerked the rifle. "Over there. With them."

Before he could do anything stupid, Nalek tugged Kol over to where we stood. Renor gave them an oddly exasperated look.

"Jory." Nalek scratched the ridges on the side of his dark neck. "Why is the Ramothorran pointing a weapon at us?"

"Because we just found out this deceitful sack of waste isn't taking us to Reytalon," I said, and spat on the deck. "He's programmed the ship to go to Garnot."

Kol swore in Jorenian. "I knew this would become a farce."

Nalek looked puzzled. "What is on Garnot?"

"A bunch of flesh peddlers who until recently have been passing themselves off as an open artist colony, to get fresh slaves," I said. "We'll have to take over the ship." I showed the Ramothorran some teeth. "When we do, I get first shot at him."

"No shot." Uzlac showed me all the gaps in his grin. "No Reytalon. Garnot." He made another gesture with the rifle. "You go quarters now."

On the forced march back to our deck, I got close enough to volley one last promise at Uzlac. "I'm going to rip your belly open, you miserable scum-sucker, and make your insides into a necklace for you."

"You make me big credits," he said, and chuckled as we filed into our quarters.

The girls and Osrea were waiting for us.

"We can't be there by now," Danea said, then went silent as she met Renor's gaze. "No. *No.*"

They had some kind of telepathy going, I realized. "Yep." I looked at the others. "Go ahead and yell at me; I'm the one who hired him."

Kol picked up something heavy and threw it across the room. It crashed into one of the sleeping platforms and totaled it.

Osrea looked interested in doing the same, if a little perplexed as to why. "Why would we yell at you, Jory?"

"For getting us into this mess." I sat down and rested my chin in my hands.

"The Ramothorran is not transporting us to Reytalon," I heard Renor say. "He is taking us to Garnot."

Kol threw something else. More crashing sounds.

Osrea still didn't get it. He'd spent too many years in his hole. "Why does he take us to Garnot?"

"It is a slaver depot world, Jory says," Nalek said. "We will be sold there."

Galena collapsed on a sleeping platform and started crying with little squeaks and gasps.

CHAPTER EIGHT

"The traveler who seeks only the path's end has yet to find the beginning."

—Tarek Varena, ClanJoren

Uzlac must have planned his little double-cross far in advance, because our room controls were inoperable and all power that would have normally routed to them had been cut off. I found this out when I tried to manually open the door panel.

"Damn it." I slammed the access hatch shut and banged my head into the wall a few times for good measure. "I can't override anything. We're stuck in here."

"So we do nothing?" Danea walked up to me, her yellow hair bristling. No, not bristling, *writhing*. "How convenient for you, Terran. Perhaps you have arranged that we remain 'stuck in here.' "

"What are you? Compulsively neurotic?" She was really starting to get on my nerves. "I didn't even want you guys to come with me, remember?"

Danea's purple lips thinned. I could feel static waves rolling from her now. No wonder her hair was acting like a nest of snakes. "How are we to know you are not the Ramothorran's cohort in this? That you did not seize the opportunity when we decided to come with you?"

"Me and a Ramothorran? Please." I waved a hand under my nose. "Credit me with better taste. And a working sense of smell."

"Do you know what slavers will do? Especially with the females?" Osrea came at me. If he'd had hair, it would have been standing on end. Instead, his black serpentine tongue stabbed the air in front of my face. "Did you make an agreement with that Houseless scum to sell us, Jory?"

Suspicion bloomed on every face around the room. Here I'd thought I'd won at least partial trust from them. Stupid me.

"Do you think I'd get even marginal credits for your big mouth?" Because it hurt, I took pleasure in hitting back. "Look at you. All of you. Mongrel crossbreeds with no education, no training, no value whatsoever, even to your own people. So what are you good for? What *could* I get for selling you?"

I paced around each one of them, making a pretense of evaluating them.

"With your handy limbs, Snake Boy, you'd go to a processing plant of some kind. Spend the rest of your life chained to a line platform, sorting components." I looked at Nalek. "Large and Dark Green here would be snatched up by a mine manager, and never see the stars again. Kol, too, unless some aristo's wife craved a big bed toy. Renor, well, the scientists would probably want to chip away at his hide for a few decades, or a jeweler would prop him in a display window. Danea and Galena would end up on their backs, servicing whoever wanted something a bit exotic."

"And you?" Osrea was just about snarling. "What brothel owner would want you?"

"Unless they cut out my tongue and my teeth, none," I said, and laughed. "I'm much more valuable on the open market—after all, I'm a pro runback with a famous name. Where I come from, people worship me like a goddess. They'd play me in some arena game. Laser-mark. Blast rallies. Tyro-shockball."

"So you would play games, while we are abused and worked to death." Danea made a disgusted sound. "Why does this not astonish me?"

"You'd probably survive for a few years, Sparky. If they cut out *your* tongue. I wouldn't last half as long." I jerked up the leg of my trousers and unfastened the thermal wrap around my knee. Time for everyone to grow up. "Not with this."

Nalek came over and bent down, peering at the mass of twisted scars around the silvery joint grips. Gingerly he touched one bracket. "It is cybertech. The kind the League uses to build drones."

"Give the Jorenian a cigar." I tugged the thermal closed and dropped my trouser leg.

"Why would they do this to you?"

"After all the fractures I received playing shockball, there wasn't enough bone and muscle left to hold my leg together."

"You should have stopped playing," Galena said in a soft, mournful little voice.

That I should have.

"But she did not." Danea had the least amount of sympathy for me, judging by her expression. "She went on playing that ridiculous game, injuring herself, year after year—"

"Eight months."

She took a step back. "What say you?"

"That's how long I played before I had to get the bootleg 'botleg. Twenty-two fractures in eight months. Nobody likes rookies in shockball, so they slam them in every game. If you're already injured, they go for whatever's bandaged. If I hadn't gotten the replacement, I'd have been a cripple for life." I let that sink in for a minute. Suddenly everyone got interested in the wall panels. "You're right, you know. It was a stupid game, and I kept playing it voluntarily. But at least I had enough spine to try. You guys would end up crying for your ClanMommy after one quarter."

Jorenians hate it when you call them cowards. I could tell how they felt by watching them. Danea wanted to fry me. Osrea wanted to pound me. The others—except for Renor—looked like they were contemplating variations on those themes.

Only Renor sat off to one side, concentrating intently on one of the bare walls.

Then something rather amazing happened. Kol stepped in front of me and shielded me with his bulk. "Sajora did not betray us to the Ramothorran."

"What say you, Kol?" That was Nalek, and even he sounded pretty upset. "What proof have you?"

"She had no reason to do so," was all Kol said.

It gave me a nice warm feeling, him standing up to defend me, but I was used to handling stuff on my own. I prodded his back with one finger. "I'll fight my own battles, thank you. Step aside."

He didn't. Kol swung around and bent until our noses nearly touched. One of his hands encircled my throat and pressed against the flat scar beneath it. He appeared ready to throttle me, so what he said next really startled me. "You are my Clan-Sister. I defend my kin."

I was still digesting that when he turned his head and repeated the same to the others. Slowly everyone backed off, including Danea.

"Thanks." I covered his hand with mine. "You can let go now."

He leaned forward instead and placed his mouth next to my ear. That rain-and-pine smell of his tickled my nose. "If I learn you have betrayed us, I will make you my first kill."

Always nice to be someone's first something. I swallowed against a dry throat and nodded. Only then did he let go.

No one wanted to chat after that, so I spent my time trying to figure a way to get out of the room and go after Uzlac. There were no access ports behind the console units wide enough to allow even skinny little Galena to crawl through them, even if I could convince her to do so. Nothing that was strong enough to pry the door panel open. And no substance corrosive enough to eat through the plasteel walls.

"Danea." She gave me a hostile stare. "Any chance you could use some of that biojuice you're producing to jolt the panel controls back into operation?"

"No." The crackling tension around her increased until we all felt the little jags of energy brush against us. "My corporeal field only affects living flesh."

I made a face. *Bet no one cuddles up with her very often.* I asked Renor if he had any ideas, but he only gave me a blank look and went back to staring at the wall.

Great. Plas-Facc was starting to lose it, too.

I went over the dilemma in my head a few times. Uzlac must have done this to more than a few passengers, I thought. They would have tried to escape, so he'd have fortified these rooms into a cage. Was there something he'd missed, something we could use?

It would take a few days to get to Garnot—I hoped—so I had some time to figure it out. If the claustrophobia and the kids didn't drive me nuts in the meantime. My stomach growled, and I rubbed it absently. Maybe I should get something to eat, keep my mind off how irritable I felt. I was already starting to sweat, despite the regulated temperature of the rooms.

Temperature.

"I'm hot," I said out loud, and started checking the upper deck surface over our heads. "Anyone else hot?"

No one answered me.

"He'd conceal it if it was up there, wouldn't he?" I looked around, picked up one of the chairs Kol had tossed, and stood it upright. After I climbed on top of it, I ran my palms over the decking, one section at a time.

Kol came over to stand beside the chair. "What are you doing?"

"Trying to find something."

"What is it?"

He sounded like he wanted to throw some more furnishings around, so I stopped and gave him my full attention. "There's a supply air duct over by the prep unit. See it?" He glanced back at the tiny six-inch port, then back at me. "Where there's a supply, there's a return."

Kol watched. It took an hour before I found the screening, which was covered with a porous material the same color as the decking. I had to use my claws to pry it off, but by then they were extruded anyway. The return air duct was about twenty-eight inches square, and went straight up.

"Nalek." I climbed down off the chair and started shedding my clothes. "I need a hand over here."

Kol immediately got pushy. "I will assist you."

"I don't think so." I gave him the once-over. "No offense, but I weigh a ton, and Nalek has more muscle." Plus I didn't want him touching me while I was naked. At least, not in a room filled with other people.

"You are going to *climb* into that little hole?" Galena asked, squeaking on the last word.

"Yeah, I am." I left my thermals on and kicked off my footgear. "Nalek? Sometime today, if you don't mind?"

The big guy had been off in one corner the whole time, being his usual placid self. Now he walked over and sized up the situation, then shook his head. "You will get stuck in there."

I thought of all the shafts Rijor had pushed me through when we were kids. Usually with both hands and yelling at me not to freeze up or he'd pound me. "I've squeezed through tighter spots than this one, believe me."

"No, Jory, I will go." Galena stood by Nalek, her winglets fluttering with quick little jerks as she looked up.

"I'm guessing you don't like small spaces, Birdie," I said.

Her chin lifted. "It matters not. I will fit better."

"What happens if Uzlac finds you? Remember how cute he thinks you are." I wasn't letting her do this. I was tough; I could take whatever the Ramothorran dished out. She'd crumple like a defender under a five-man line drive.

"I will not allow him to catch me." Little Galena started to strip. Unlike me, she didn't have thermals, so she went down to her skin. Which was when we all found out she didn't have mammary glands or body hair. "Tell me what I must do."

If she was that determined, well, she'd make it through faster than I could. "Okay, first, crawl through the shaft."

It had been more than an hour since Galena disappeared into the return duct. Everyone moved restlessly around the rooms. Everyone except Renor, who had stopped staring at the wall and now watched the hatch with the same intensity.

Osrea was the worst. Snake Boy had paced a continuous circle from the chair under the duct to the door and back again. He kept grinding the palms of his hands together, and the hard blue exocartilage plates covering them made a *screech, screech* sound.

"Where is she?" he asked me. Again. "Could she be trapped? Why has she not yet attempted to release the door panel?"

Osrea had started getting on everyone's nerves a long time ago. We were now, as a group, ready to kill him.

"Os," I said, rubbing a hand over my eyes. "Go sit down and shut up, and stop grinding your plates together. Or I'll hurt you."

He didn't take my advice, but came over and grabbed my arm. "Did you do this to make it easier for the Ramothorran to put his filthy hands on her?"

Kol instantly got to his feet and came at me from the other side. I waved him off.

"Listen, Os." I kept my tone low and sympathetic. "I know you're worried about her. We're all worried. But give the kid a little more time. Birdie's not like you. She's scared to death of enclosed places."

"Knowing that, you let her do it? You let her go in there?" For a lizard, Osrea could certainly be hot-blooded.

"Take it easy, Snake Boy. It's probably the first time in Birdie's life that she's been allowed to do something reckless."

He started to say something, then changed his mind when Kol glared. "Do not call her Birdie. I do not like it."

"Fine." I smiled. "Stop crushing my arm or Kol is going to rip your throat out."

That was when I heard someone fumbling outside the door panel. We looked at each other and practically ran over to it. A moment later the panel slid slowly to one side, and an extremely dirty but grinning Galena stepped inside.

"I became wedged in corners a few times, but something seemed to push me along and"—she made a broad, triumphant gesture—"I made it through."

"Way to go, sweetheart." I gave her a one-armed hug, then put her garments in her hands and peered around her. "Get dressed; then let's move. And for Pete's sake, everyone be quiet until we locate Uzlac and disarm him, okay?"

Osrea looked mystified. "Who is Pete?"

"Never mind. Come on."

* * *

We crept through deck after deck, searching room after room. We found plenty of filth. Clutter. Malfunctioning or inoperable equipment. Some very suggestive female garments. Illegal substances. Even some weird-looking fungus growths sprouting in the strangest places. Yet after a thorough aft-to-stern sweep, even I was ready to admit defeat.

No Uzlac. Anywhere.

"He wouldn't have left the ship," I said as I stalked around the helm, then paused and glanced at Renor, who was quietly reprogramming the navigational array. "Would he?"

Renor simply shook his head.

"He's too fat to hide with any degree of success." I thought about it for a minute. "Could he have accidentally blown himself out of a pressure lock? Could we be that lucky?"

Everyone was busy doing something and ignored my rambling. But Renor, I noticed, sat just a little stiffer in his seat.

What Galena had said came back to me: *Something seemed to push me along.* Something had kept Danea from tearing my throat out back in the caves and in our locked quarters, too. Maybe it hadn't been an accident.

Maybe Uzlac had gotten some assistance to the nearest exit. Announcing that wouldn't be very diplomatic, so I went over to the array and sat down next to Renor.

I nudged him. "You want to get something off your chest?" He looked down at himself, puzzled. "I mean, do you want to tell me what you really did to Uzlac?"

"No." He returned to inputting the new course settings.

"Want to at least tell me if he's still breathing?"

Renor's eye slits contracted to twin vertical lines. "Not where he is now."

"Good." I didn't want to push him, but there were some things I *had* to know. "Danea's jolts only work on living beings. I assume whatever killed Uzlac has similar limitations." I got a tiny nod. "Shame whatever killed Uzlac didn't do it a little earlier, like when he pulled the gun on us."

"Fear of discovery often outweighs opportunity."

Then it all clicked. "People might want to lock up someone

who could push a living being around with a thought, or whatever." He didn't twitch a single surface. "Okay, Ren, you can keep your secrets. I will, too."

It took Renor some time to figure out Uzlac's shoddy programming and reroute the ship to Reytalon. Turned out the greedy slob had encoded the pertinent data with a bunch of nonsensical algorithms, in the event the ship's systems were later impounded and examined by Jorenian officials. He'd set up a few data-wipers, too, which were a real headache to disable.

That was one thing slavers were good at—ditching the evidence.

Everyone hung out at the helm for a few minutes, crowding each other and making a general nuisance of themselves. Only when Osrea accidentally bumped one of his arms into a console and disabled the aft scanner did Kol abandon whatever he was doing with the com console to take charge.

"Nalek, you and Galena attempt to secure individual quarters for each of us. Osrea, Danea, check ship's stores and assure we have enough provisions to complete our journey. That Ramothorran may have consumed everything edible on the vessel." Kol turned to Renor. "How long until we reach Reytalon?"

"Approximately six point two rotations," Renor told him.

"When the four of you have completed your assignments, return to the helm and report to me." Kol made an abrupt gesture, and the other Jorenians took off. Renor was still engrossed in his computations.

That left me, and my ClanBrother gave me a long, thoughtful look. "Sajora."

"I don't clean lavatories," I said at once. "Or viewports. Prep units hate me. So do sanitizers."

"Do you know anything about weapons systems?"

"Only what I learned from the underground, which was mostly Terran, salvaged off junked ships." I frowned. "Why?"

"Uzlac signaled Garnot immediately after we left Joren." Kol waved me over to the communications array. "When we do not arrive on schedule, they may send a vessel to track us down."

"How do you know they . . ." I got distracted by a relay Kol had punched up on the console, read it, then swore. It was an invoice of delivery. "That bloated little toad got paid in advance." A hefty chunk of credits, too. "And after I poured every chip to my name into his greasy palms, the leech."

"Ramothorrans are not known to be particularly ethical," Renor said in an absent tone.

"No kidding." I elbowed Kol aside and pulled up the local star charts. "We're only two light-years from Garnot. For this much money, they'll *definitely* send a ship to recover us." I swiveled toward Renor. "We've got to get the hell out of here."

"As you said, no kidding." Renor completed his input and rose from the console to activate another. "Kol, as you requested, here are the security grid controls and a current inventory of defensive armament. As I have no experience in this area, I am unable to advise you if it is adequate for defense purposes. However, the majority of the inventory is of standard Terran manufacture."

Both of them looked at me.

"Whoa, hold it right there." I showed them some palm. "I didn't have anything to do with weaponry on this ship. Having a little DNA and living on the planet doesn't mean everything Terran is my fault."

"You said you learned about Terran weapons in the underground."

"Yeah, so?"

"You know more about it than we do." Kol took my arm and guided me over, then pushed me into the console seat. "Tell us what you can."

I studied the screen. Four standard pulse emitters, all drained, two launching shafts, and a couple dozen of the oldest series of displacer projection torpedoes in existence. "Oh, great. No juice in the emitters. We used to steal protorps like these from military recyclers so we could blow new shafts."

A big hand landed on my shoulder. "Is it enough to defend the ship?"

I reconsidered the list. "Against one Garnotan in a very thin envirosuit, maybe. A whole ship of them?" I shook my head. "We don't have a prayer."

"Then we should, as you say, get out of here. See to it, Renor." Kol grabbed my arm and tugged me out of the seat. "Jory and I have other business to attend to."

I didn't care much for the way Kol was bossing everyone around. Now this hauling me from here to there. I planted my feet. "What business?"

"Family business." He pointed to the chamber Uzlac had occupied while the ship was on autodrone, just off the helm. "In there."

Either Uzlac had a fascinating collection of etchings, or Kol wanted to yell at me. I was guessing the latter. *Okay, maybe he's entitled.* Silently I went into the dark, smelly room and activated the light panel. And immediately wished I hadn't.

"God, what a pigsty." Rotting food, stained garments, and anonymous filth lay piled all around us. I'd seen arena lavatories after ten thousand fans had gotten through using them. They'd been cleaner than this.

Behind me, Kol closed the door panel. I faced him, saw the look in his eyes and the way he was standing. "What?"

"Jory, you came from Terra, a world of dedicated xenophobes, to Joren, a world from which your ClanMother was banished." He rolled his hand over the back of his neck.

I took a wary step backward. "Uh-huh."

"You traveled across Joren to gather us together and tell us we are the progeny of slavers, born outside bond."

"We are."

"You tell us you intend to become an assassin, so you may seek out and kill the raider who was responsible for the dishonor of your ClanMother." That was when his claws emerged. "And the dishonor of all of our ClanMothers."

"Right." I took another step back. Why was he so angry? "Kol?"

"You contracted space for this journey with a trader who had already sold us to slavers before we came on board. The only weapons he has stocked also happen to be Terran." He'd been leaning against the panel, but now he started walking toward me. "How did you phrase it? *Do the math*."

He was right: The facts made me look terrible. Terribly guilty, or terribly stupid. Kol was about to decide which.

I could have looked around for something large and heavy to club him over the head with. I could have dodged him, maybe long enough to get out of the chamber. I could have yelled at him for being so damn stupid as to think I'd go through all this waste just to snag Uzlac a half dozen exotic-looking Jorenian slaves. Only I was trying to figure out how Uzlac had known I was bringing six people with me—and had sold them—before I ever said a word to him.

He stopped, white-within-white eyes hot now. One broad hand lifted and extended until his claws were only an inch from my face. The smell of rain and pine was so strong it filled my head.

I focused on those five gleaming, dark blue points. One time Mom caught me trying to saw off the sixth finger on my right hand, so I could look like the other Terran kids. Kol, who had lived on the planet I should have been born on, had also gotten the ten fingers I should have been born with.

"I didn't arrange this. I didn't plan to bring any of you with me. If Uzlac signaled Garnot from Joren, maybe he did it right before we took off. Or he had someone watching us. I don't know."

"Indeed." My ClanBrother's deep voice dropped to something that sounded like it came from a black pit. "I said if you betrayed us, I would make you my first kill."

He could try. I kept my head back, stared him directly in the eyes, and didn't twitch a muscle. "I haven't betrayed you. I know how bad it looks, but I'm telling the truth."

The claws didn't retract. "Swear it to me," he said, so low it was barely a whisper. "Swear it to me as my ClanSister."

I knew how much danger I was in; I could see the tremors of violence twisting in his muscles, just beneath his skin. Sweat trickled down the side of my face as my own inner beast reared its head. "I swear to you, ClanBrother, by the ties of blood and House, *I have not betrayed you or my kin*."

His hand dropped, but Kol was still enraged. I was equally furious. There was also this very peculiar sensation that made me resent the scant space between us. I didn't know why, or what it was, but something had to give. When it did, I'd bet neither of us would come out of it undamaged.

"Unless you want to end up as my next kill," I said, equally low and soft, "walk away now."

And to my surprise—or disappointment—that's exactly what he did.

CHAPTER NINE

"Blessed is the traveler whose path chooses him."

—Tarek Varena, ClanJoren

Kol went back to the helm. I decided to head down to the launch bay so I could check out the shaft tubes and scan the inventory. And, in the process, stay as far away from Mr. Faith and Trust as possible.

There were two things that got me so aroused that my claws emerged. One was rage, and that I had plenty of. The other was sex.

I remembered Rijor, laughing. *If you tear another hole in my tunic, I am making you sew it up.*

I hadn't taken a lover since Rij died. I'd never gone in for casual sex, anyway—it would have been suicidal, trying to play ball on a mostly male team with that kind of rep—but I missed it sometimes.

Apparently I'm missing it more than I thought. And Kol's not helping.

An hour later I was still angry at Kol, still not sure what to do about the other feelings I was having, and still calling Uzlac every filthy name I could think of.

A third of the protorps were so old their warheads registered total flat-line—which meant they were inert, useless hunks of corroded alloy. Half of what *was* usable had been slowly leaking radiation over time, which in turn contaminated most of the storage compartment. I was forced to put on a reinforced envirosuit before I could finish my inspection.

"Who keeps junk until it acquires a half-life of ten thousand years?" I shouted at no one in particular before I went to the nearest panel and signaled the helm. "I'm going to need some help down here before we have a nuclear incident."

"I will send Danea and Osrea to you," Kol replied, sounding just as annoyed.

"Send them soon."

The only good thing was that the units were the same old Terran models we'd used a couple of times to tunnel out new passages when the PRC had collapsed the underground's old subway network, and I was comfortable using them. I figured out the rest by studying the ship's specs off the database and using common sense.

Osrea and Danea finally appeared, neither of them looking like my best friend, but no claws showing. I guessed Kol hadn't told them about the Ramothorran's presale yet.

Either he believed me, or one of us was going to end up decorating the wall panels with the other's entrails. I didn't want to dwell on why I suddenly cared so much about what Kol thought. It wasn't like either of us could jump ship to get away from the other.

"About time you showed up." I wiped my dirty face on my sleeve and eyed the last of the shafts I'd checked. Uzlac had left a protorp in the tube, but it was just another dud. "Anyone want to buy a large, useless object?"

Danea's purple nose crinkled. "You would sell yourself, Terran?"

In spite of Kol, in spite of Uzlac, I had to laugh. "Good one, Sparky."

"Kol sent us to assist you." Osrea jumped down on the loading platform and peered into the shaft. "Are we placing the missiles in there?"

I shook my head and pointed to the grav-hoist. "The hauler will take care of that. We've just got to move them out of storage and over to the platform." I had found a semifunctional liftrig, but I'd never used one before. "I need Nalek down here."

Snake Boy got indignant. "I can do whatever you require of Nalek."

"Sure, pal." I dragged him over to the rig, which was, like everything else on Uzlac's ship, falling apart, rusty, and sprouting wires at every coupling. "Climb in."

Danea had trailed after us, and inspected the rig. "He will be harmed if he uses this equipment."

That was enough for Osrea. The next thing I knew he was clambering up into the body cab and working his limbs through the harnesses. He was too wide and not tall enough to be comfortable, but eventually he got himself wedged in sideways, strapped up, and activated the control unit.

I told him where to move the hauler, then glanced over at Danea, who was doing her bad-hair-day thing again. "The guy is just a masochist, isn't he?"

She studied her fingernails, which matched her skin. "If he is injured, I will see you suffer for it."

Nice thing about Sparky, you always knew where you stood with *her*. "Fine. If you want to keep the snaky hair, put on an envirosuit."

Machinery whined as Osrea initiated the rig's forward grapplers and worked them up and down for a minute. Gaskets popped. Wiring arced. Metal groaned. The noise was only slightly worse than the smell. I figured the rig would hold together while we loaded the shafts, but no longer. If there was a God.

Os started moving in the wrong direction.

"You're in reverse!" I had to shout to be heard over the grinding gear shaft. I waved toward the storage compartment. "That way!"

With some difficulty Osrea maneuvered the ungainly unit from the place it had been rusting into the deck over toward the compartment. I hurried ahead, but Danea got there before me, and opened the panel. I'd pointed to where I'd marked the inert units, then indicated the ones we needed Os to retrieve.

"He can't squeeze the entire rig in here, so we'll have to help him guide the grapplers," I said, using the suit comlink. "Stand by the panel, and use hand signals to relay what I say to him."

It took a few tries before we were able to steer Osrea's grappler arms to the proper position. I adjusted the end clamps myself, then made a circling gesture. "That's it; tell him to retract them."

Osrea pulled the protorp completely out of the compartment and up into the buttress, after which we edged out of storage and took up positions on either side of the rig.

"Watch those clamps," I said to Danea. "One of these things falls, we'll all glow in the dark for a few centuries."

We flanked Osrea as he went into reverse, turned, and drove toward the launching tubes. The additional weight didn't make the trip any faster, and the rig made even more noise. Once we reached the loading platform, I initiated the grav-hoist program, then climbed down beside the tubes. Danea and I used the same hand-signal relays to guide Osrea, and hoisted the protorp down onto the tube conveyer.

"The rig will not hold up much longer!" Osrea shouted from the cab once we were through. "We must haul two at a time!"

It was tedious, sweaty work, back and forth and up and down, over and over and over, but we kept at it. Transporting two units at the same time severely taxed the buttress stabilizers, and did nothing to make the grav-hoist any steadier. At last we had the final pair of protorps out of storage and in position over the platform.

"Nice and easy," I told Danea after we'd hoisted one down. One of the buttress struts started to loosen from its base. "Tell him to take his—"

The ship suddenly shuddered, then pitched violently to starboard. I ended up sprawled across one of the units, while above me Danea was thrown to the deck. From my position I could see the bottom of the rig, and held my breath as it tilted ominously up in the air. A second passed, then two, before it finally righted itself and hit the deck with a huge thud.

The last protorp—luckily—didn't immediately drop out of the grapplers and on top of me.

Danea got to her hands and knees, looked over the edge of the platform, and shouted, "Get out of the way!"

The feedback her yelling created on my suitcom nearly punctured my eardrums. I tried to push myself off the unit, but the ever-present filth on everything had made the surface of my suit slippery, and I lost my grip. Above me, one of the stabilizers blew, which made the buttress collapse. Instantly the right grappler dropped out of control. The protorp falling on top of me weighed about four hundred kilos, so I didn't even bother to pray. I simply wrenched my body sideways and tried to roll off.

A blast smashed into my face and propelled me off the protorp, and over the side, and slammed me into the conveyer slot. From the way my eyes bulged and my limbs jerked, it felt like a triple-penalty charge.

That couldn't be right, I thought, just as something really big and heavy landed a few inches away. The resounding crash made my ears ring and blood well up in my mouth. I'd never gotten any third penalties. *Well, not counting that lousy call down in Florida . . .*

"Jory!"

One of my teammates pulled me up on my knees. I gazed at her through her faceplate (when had the junta agreed to going full plas?), sure she knew why I'd been clobbered so unfairly.

"Got it? *Got 'em?*" My head hurt; why the hell did my head hurt; where was my helmet? And why was she fooling with me? "Fourth and goal, forget about me, shake your ass!"

Behind the helmet, yellow hair was snaking all over her eyes and mouth. Stupid bitch, why'd she wear it loose? *Must be a greenie.* Rijor was going to eat her for breakfast.

Rijor's dead, idiot. You watched him die at the Ditka Dome, remember?

The brainless rookie wouldn't let go. She dragged me up and out of the hole, threw me flat on my back, and straddled me. Then she was looking up and shouting.

She'd blown the whole play, plus no scanner in the world was going to miss her little stunt. I'd just have the pleasure of watching the compref shock the shit out of her for unsportsmanlike conduct on the field.

Only we weren't on any field.

My head cleared, and I remembered. *Danea. Osrea. Uzlac's*

ship. I wasn't in New Angeles, getting myself half killed for the pleasure of screaming shockball fans. We were on the way to Reytalon and we had to take care of . . .

The protorps.

I pushed Danea away, sat up, and pulled my helmet off. Oh, my aching head. One look confirmed the units were still down in the conveyer and didn't appear to be ready to explode. Then I turned my head and nearly shrieked.

Osrea, bleeding from several bad gashes in various places, held out three of his hands. "Are you well, ClanSister?"

"Just great." I let him haul me up. When Danea reached to steady me, I jerked out of reach. "Thanks, Sparky, but a blast from you I don't need right now."

Danea muttered something vile and stalked off.

"Damn." I rubbed my head. "What hit me down there?" I glanced at the rig. "Os, how did you do it? That stabilizer is fried."

"I did nothing." Osrea jutted his chin toward the glowing, retreating form. "Danea projected her . . . field?" He shrugged. "It knocked you off the missile before the other dropped on it."

Terrific. Now I owed the bad-tempered witch for saving my life. "Let's get this junker moved; then I want to chat with the idiot who's flying this crate."

Osrea and I hoisted the fallen protorp into place, then stowed what was left of the rig. After I cleaned him up with a first-aid kit, I left Snake Boy to man the bay.

"Don't—I repeat, *do not*—press any console buttons," I said as I went to the lift. "I'll signal you when I find out what's going on."

"Remember to thank Danea for maintaining your path," Osrea said, stretching all four upper limbs with a grunt.

Was he smirking at me? "Yeah. I'll be sure and do that."

Back up at the helm, Nalek, Renor, and Kol were stationed in front of various consoles, and something bright was filling up half the central viewer.

I didn't bother with pleasantries. "Now is not the time to be joyriding, boys. I nearly got flattened by a protorp down in launch bay."

"We are not riding joyfully." Nalek turned his head for a moment. "We're under attack."

I shut my mouth and sat down beside him. While I was strapping into my harness, I noticed he was transferring power from the stardrive to the forward emitters. "What's that big shiny thing out there?"

"I have no idea," he said. "It appeared shortly after the other ship fired on us."

I ran a scan, but all that registered was an anonymous power source of unknown origin. In the meantime, I located the other vessel—one twice our size—a thousand klicks from the shiny thing. The scanners confirmed it was fast, powerful, and absolutely Garnotan.

The slavers had come to recover their shipment.

I put a signal through to launch bay. "Osrea, it's time to start pressing buttons. Get on the sequencer and wait for my signal." By that time Nalek had transferred operational command to my console. "Nal, I'm sorry, I didn't mean to take over."

"By the Mother, please do." His teeth flashed. "I am happier performing other tasks, ClanSister, I assure you."

"Good man." I love a natural team player. "Head down to launch bay and give Os a hand."

Once Nal left, Kol transferred his console over and sat down beside me. "The slavers have altered their heading, possibly to drive us into the energy anomaly."

Was it some kind of trap? Only one way to find out. "Os, prep both tubes for launch." The forward emitters were only half-charged, but I didn't want to drain any more power from the stardrive. I glanced sideways. "Jolt them or blast them, Captain?"

That seemed to startle him. "I am not the captain of this vessel."

"I don't think anyone else wants the job. Jolt or blast?" When his expression didn't change, I sighed. "Jolt them with the emitters, or blast them with a protorp?"

His mouth curled. "Blast them."

"Osrea, you prepped?" I waited for his affirmative, then transferred the targeting data. "Lock on these coordinates, and fire tube one."

Something made the ship shudder again. I checked the screen, but the slavers hadn't fired. Then the first volley flew from our launch bay and headed directly for the Garnotans. "Os, are you okay down there?"

The sound of coughing came over the audio. "By the . . . Mother . . . when was the . . . last time . . . the Ramothorran descaled . . . these tubes?" More coughing.

"Probably never." I monitored the protorp, which reached the slaver vessel a moment later. There was a small bloom of heat, and the Garnotans came to a dead halt. "Nalek is on his way to help you. Keep prepping those tubes." I terminated the signal and ran the scanner.

Kol watched my screen. "Full impact?"

"Yeah." I gnawed at my lower lip. "Stopped them in their tracks, but didn't do any damage to the ship."

Renor called from his console. "We are receiving an incoming signal from the other vessel."

"Ramothorran vessel." An amiable-sounding voice came over the audio. "Cease fire and stand down or we will send you into the ion-well, and you will be destroyed."

"Nice guy," I said. "Very straightforward. What's an ion-well?"

"It is the big shiny thing," Kol said.

Renor made an odd sound. "It is *not* a viable alternative to capture and enslavement."

Kol opened a relay channel. "We will not stand down, slaver." He nodded toward my console.

I put through adjusted targeting coordinates and transmitted the order to fire to Os.

An instant after the protorp left the tube, the Garnotans attacked with a scattered barrage of emitter pulses. Uzlac's ship rocked and rolled, and a huge explosion sent the ship veering forward toward the ion-well.

"Damn it." I checked my screen to be sure. "They blew the protorp before it covered a hundred klicks. Osrea?" I had to repeat myself a few times before he return-signaled. "What's going on down there?"

"Tube one release controls are not responding. I think the outer hatch is fused." Osrea coughed. "Conveyers are down as well. Nalek is loading tube two manually."

Nalek was stronger than I'd thought. "Bypass what you can, and let me know when you're ready." I turned to Renor. "If you can move this heap, I suggest you start making some defensive shifts. Now."

"Helm and propulsion controls are not responding." Renor's hand sparkled as he pointed to the central viewer. More pulse fire rocked the ship. "We are being driven into the well."

Our defiant stand lasted all of about three minutes.

We got off one more protorp, but the slavers once again detonated it just outside the tube. That caused substantial damage to launch bay level, forcing Nalek and Osrea to evacuate the deck.

I drained the paltry emitter charges by firing first at the slavers, then, at Ren's suggestion, at the ion-well looming in front of us.

"The shock wave may repel the ship away from it," he told me as I fired.

A fraction of a second later, the ship rocked violently, but continued toward the well.

"Or not," I muttered as I switched targeting back to the slavers.

The Garnotans dodged nearly every volley and were largely unaffected by the two I landed.

The closer we got to the ion-well, the more its gravity latched on to us. We were being dragged in.

In the meantime, Kol had ordered everyone to grab what weapons they could find and report to the helm. Renor quietly kept trying to regain control of the ship, but the ion-well wasn't taking no for an answer.

"The helm is unresponsive." Plas-Face moved back from the console. "There is nothing more I can do."

The other crossbreeds arrived. Galena's iridescent eyes grew huge as she saw the light filling the viewer, while Danea hovered close to the nearest exit. Nalek and Osrea trudged in, both covered with grime and still coughing.

Over the audio, the pleasant voice spoke again. "Prepare to be boarded for immediate evacuation."

"Prepare to drop dead." Fury made me hit a wall panel. They knew we had no choice, and I'd bet they were chuckling about it. I went over to Ren's console. "Can you fly into this well thing before they get a launch to us?"

"We will be destroyed," Danea said, suddenly right there on Renor's other side.

"I'm not letting them run me to death in some alien arena." I gave her the once-over. "What, do you *want* to be a whore?"

Her white eyes narrowed. "We can fight them!"

"Yes, we can fight them," Kol said, "but in this situation, we cannot prevail." He got up and made an eloquent gesture. "There is a launch on board. Those of you who choose this path may take it and surrender to the Garnotans. It is possible that, in time, you may escape your enslavement."

No one moved, and something frozen inside me turned to slush.

"Slavery is extremely overrated," I said, and winked at Nal. "So is playing contact sports."

Nalek regarded his huge hands. "I've always preferred building to digging."

Galena, tears streaming down her face, gulped. "I would not make my owner many credits, I think."

"Yes, you would," Osrea said, then looked quickly at the deck. "Sorting components is boring."

Renor got to his feet. "I would prove a . . . tedious laboratory specimen."

"Think you I *wish* to be a prostitute?" Danea glared at everyone; then her hair settled down and she slumped back against the wall panel. "Oh, very well. I still say we could have given them a memorable battle."

Kol made another of those elegant bows. "I am proud to name you all my kin. Renor." He turned to the viewer. "Fly the ship into the anomaly."

Renor sent the last of our power into the stardrive, and initiated propulsion.

The nice guy from the slave ship suddenly became very agitated. "Your vessel cannot withstand the forces within the ion-well."

"We are aware of this," Renor said, and made a slight adjustment to take us into the center of the anomaly.

"You will all be destroyed."

Osrea tapped the com console. "We *know*, slaver."

"Stop immediately and we will retrieve you."

I did the final honors. "Go mate with yourself, flesh peddler."

As we watched the gleaming field expand to fill the viewer, Kol came to stand beside me. His voice was amused. "Surely you could have offered something more insulting than . . . flesh peddler?"

I shrugged. "Birdie doesn't need to hear that kind of language." I didn't object when one of his long arms settled around my waist. I think I even leaned my head back against his shoulder. In formal Jorenian, I said, "Your pardon, warrior."

He rested his cheek against the top of my head. "No pardon is required, lady."

I breathed deeply, wanting to remember his scent. This close to him, it was like walking through pine trees in the middle of a storm.

Renor lifted his hands from the console. "Impact in ten seconds."

When the ship entered the outer perimeter of the ion-well, the subsequent jolt sent everyone flying to land on the deck, and shut down all internal power systems, so I didn't get to see what happened. In the dark I got as far as my knees before an excruciating force squeezed the breath from my lungs. The last thing I remembered thinking was, *Hull breach*.

Then I died.

Sometime later, I opened my eyes and discovered I was still alive. Alive, but naked, in the dark, and strapped down to a hard, cold sheet of some kind of metal.

My reaction was predictable. "Christ, I'm not dead." I lifted my head about an inch, but still saw nothing. The old scratches across my chest hurt, too. Must have bumped them or something. Or something? "I'm not dead, right?"

"You are alive, ClanSister."

That was Nalek's voice. I tried to lift my head higher, but the strap across my brow made it impossible. "Nal? Big guy? That you?"

"Yes, I am here." The smooth baritone sounded a little strained. "Jory, do you know where we are?"

"Not in the ion-well." If this was heaven, Rijor had gotten it all wrong. "Otherwise, I don't have a clue. Are the others around?"

A faint yellow glow suddenly flared up off to my left. "What have you done to us, you stupid Terran bitch?"

"Sparky's still around," I said, then assumed a conversational tone. "I haven't done anything yet, you cranky, foul-mouthed power outlet. But I'll keep you updated."

Groans and gasps began to erupt all around me. After a few minutes of confusion, Nalek performed a roll call that verified we were all indeed present and accounted for.

No one knew what had happened; hitting the ion-well had knocked us all unconscious at the same time.

"I thought the ion-well was a bad thing," I said to Ren. "We're supposed to be hugging the stars or whatever now."

"It is, and we are."

A strange sound made us shut up. There was a metallic *snick,* an unseen panel *shooshed;* then a low hum rippled over my eardrums. It got louder. And closer.

"What is tha—"

It hit my chest and began slicing into me before I could form another word. The explosion of raw, tearing pain made me arch up as a scream ripped from my throat. I fought it, but the pressure increased until it punched through my skin and the muscles below it. Some kind of cutter . . . rotating in a circle . . .

Pain became unbearable agony. More darkness.

I didn't *want* to wake up the next time, but someone's foot landed in my ribs. Hard.

I grunted, clutched my side, and automatically curled into a ball. And got kicked in the back for my efforts. The old reflexes snapped into play; I coiled over, got my feet under me, and pushed off my forearms.

Standing up was a real treat, considering that I did it with legs that felt as if both knees had been removed. Three problems became evident: I was still naked. Unfriendly alien faces surrounded me. I'd been dumped in what appeared to be an enormous prison cell.

I gaped at my strange cell mates. "Where am I?"

No one answered. A meteor-sized fist swung out toward my face, making me duck and swear. I swiftly backed up until my back hit bare stone. Looking down, I discovered a weird-looking bandage covered most of my right breast.

"How did I get hurt?" I scanned the cell. "And, um, arrested?"

No one could or cared to speak my lingo. The ring of uglies stretched five mugs wide, all different species of humanoids, and one oozing, dripping thing I didn't even *want* to try to identify. Not one of them appeared particularly fond of Terrans or Jorenians. They started forward, clamoring in different languages, hands and claws and tentacles extended. Not for a handshake, either.

"*Rhe avele!*" someone shouted in a high, piercing voice, and the uglies broke and ran. A very large, even uglier alien walked up but stopped about a foot away. "*Owkn mia nic rhgea?*"

"Nice to meet you too." I jerked my head around. No sign of my family. An unfamiliar sensation of being exposed made me frown. I was no prude. "Um, you speak Terran? *Locega Jorenhai?*"

"Speak some," he said in broken Jorenian. "You new. I Truk."

Was that his name, race, or occupation? I'd go for name.

"Greetings, Honored Truk," I said, and bowed as if to a Ruling Council member. I guess he liked that, from the way he snarled. "I am Jory Rask of Joren. Know you my people?"

"Joren. Blue flyers. Sell good." Truk nodded his approval, then inspected me. "You not blue."

"Only on the inside." We were a hot ticket item. Marvelous. "My people who came with me? Are they here, Honored One?"

He made an affirmative gesture, and hitched one of his scarred appendages back over his shoulder. "There. All."

I wasn't going to twitch a toe until I had this guy completely on my side. The bizarre, unidentifiable sensation of vulnerability—was I turning *into* a prude?—increased. "May I speak with them?"

He thought this over for a minute, then simply grabbed one of my arms and dragged me away from the wall. From the sudden change in the fit of his ragged, filthy garment, he had other things on his mind. "You Truk female."

"She is *my* female," I heard Kol say.

I was dragged around when the big alien swiveled to face Kol, who was standing in a deep pool of shadow. Both guys were about the same size, but Truk didn't have much in the way of claws, and Kol had all of his ready and waiting. This seemed to impress the big alien.

I was distracted by what I could see of Kol. Like me, he was also completely naked.

"You female." Tuk made a sound that might have been laughter, and shoved me across the space toward Kol. My Clan-Brother caught me with one arm. "Protect."

"Hi, Kol." My face smashed into his chest, and I found out he had the same bandaging I did. "Mind telling me where we are?"

"Renor believes it to be a slave holding pit." Kol backed away from Truk, still holding me against him. "The others are over here. Danea has been injured."

As we crossed the open cell area, I saw about thirty other species among the prisoners, making up altogether over a hundred individuals in the pit. Some were naked; others had wrapped parts of themselves with scraps of cloth. Odd thing, none of them had hair—even the ones that were supposed to.

That was when I realized what was bugging me and grabbed my head with both hands. My completely *bald* head.

"My *hair!* What happened to my hair?"

"They have removed it." Kol yanked me back into a corner, where the other five Jorenians were huddled. I reached up and swiped at his head, but found only bare scalp. "Mine and the others, as well. They have also denuded Galena's wings. Doubtless to mark us as new captives, or to control parasitic infestation."

"I'd rather have bugs in my hair. What happened? Did the Garnotans drag us out of the well after we lost consciousness?"

He glanced at the other prisoners. "From our present circumstances, it would seem so."

I moved away from him and surveyed the group. Since Galena, Kol, Danea, and I were the only ones who *had* hair to start with, it wasn't a shocking change. The nudity I ignored—I'd spent too many months in a locker room to be embarrassed by a bunch of body parts. Though I couldn't help but spot some interesting variations all around.

That's when I noticed Sparky's condition. "What did they do to her?"

Nalek looked up from where he knelt beside Danea. "I do not know, Jory, but she is very ill."

Delirious, from the sound of her moaning. I bent over and held a hand just above her brow. Her corporeal field was inactive or drained, so I touched her. The skin under my palm felt dry and brittle. "She's burning up. We need to get some water. Ask Large and In Charge over there if he can get us some."

"I will do it." Renor moved from his position by the wall and went after Truk.

Nalek gently lifted Sparky up so I could check her for wounds. Under the bandage over her breast, I discovered a nearly healed surgical scar. When I carefully pressed my palm against it, I felt a round, flat object just under the skin. "There's an implant of some kind in there." I tore the bandage from my chest, and found an identical, closed incision, and the same hard shape beneath it. "Do the rest of you have these?"

Everyone checked. Everyone did.

I looked into her eyes, ears, and mouth, but saw no sign of bleeding. I couldn't find anything else wrong with her. "Did she go after someone, get smacked anywhere?"

"No." Nalek shifted her limp form in his arms, then bent his head down closer to hers. "Jory, she is having trouble breathing."

CHAPTER TEN

"GUIDE OTHERS AS YOU WOULD BE GUIDED."

—TAREK VARENA, CLANJOREN

I was no medic, but I knew we had to get her cooled off. "Hey, Plas-Face! Get that water over here, pronto!"

There was the sound of a scuffle and some vicious swearing before Renor reappeared, his crystalline hide smeared in a couple of places with some ominous-looking fluid. In one hand he carried a bucket, which he handed over to me. "This is all there is."

I collected our discarded bandages and dipped them in the bucket. "This isn't enough, and I need more cloth." Osrea shoved something stiff, threadbare, and soiled into my hand. I held it up with two fingers and wrinkled my nose. "That the best you can do?"

Snake Boy took exception to that. "Considering we have no garments? Yes."

I took one of the soaked bandages and stroked it over Danea's brow. Renor stepped back, while Sparky made a curious keening sound, then tried to shove her whole face in the bandage. A strip of faint yellow stubble suddenly appeared where her hairline should have been. That was when I learned a lot more about Danea than she probably ever wanted me to know.

"Bring the bucket over here, Ren. How has she been keeping wet without any of us noticing her drip?" I asked as I reached for the handle.

"Under her—" Renor stopped. If his sparkling face had been

capable of expression, it would have creased with dismay. "A skin suit, worn under her clothing. How did you know?"

"My best friend was part Imabjaic—half-fish." Rijor had rigged a body seal under his garments, too. His species had to keep a thin layer of water between their sensitive derma and the elements. He'd always bitched about how his exposed face and hands had perpetually flaked.

What would you tell me to do for her, Rij? How in hell am I going to keep her alive until we can get her into some water?

I could almost hear his drawled answer. *Do whatever it takes, Green Eyes. Piss on her, if you have to.*

I would, too, if it meant saving her life—and wouldn't that thrill Sparky. For now, however, I'd use less drastic measures. I dumped all the bandages in the bucket, then took them out and arranged them over Danea's chest. She shuddered with relief.

"There. That should work for a little while, but we definitely need more water."

Kol touched my arm. "What is wrong with her, Jory?"

"Our ClanSister neglected to tell us she's an aquatic." I dribbled more water over her lower extremities, avoiding the clogged gill slits on each side of her torso. "She's simply been out of the water too long." I put the bucket down and rose to study Renor. "Either of you could have volunteered this small detail, you know. Why didn't you?"

His eye slits contracted. "It was Danea's choice to remain silent, and I was compelled to respect that."

We made some team. "Ren, without water, she'll die. I think I'd have found a chance to mention that, somewhere along the way." I looked around the filthy floor and in the bucket. "I need to clear her gills. Get more cloth so I can soak her down with what's left of the water. If we can't keep her wet, we can at least keep her damp. Ask someone if they'll lend us a loincloth, and when our keepers are going to bring more water."

Nalek was moving before I said another word, and approached one of the biggest aliens with the most rags adorning its body. A moment later he came flying through the air to land on his backside.

One big hand probed the right side of his face. "Don't ask that one anything," he said, pointing to the prisoner who had just socked him.

"All right, boys." I motioned to Kol, Os, and Ren. "Go persuade some of the smaller ones to cooperate." I rolled Danea onto her back and checked the other sides of her gill slits. All of them were swollen and a couple were oozing. "Galena. Did they miss any of your feathers?"

She turned and presented her back, and I found a small white quill stuck in where her winglets joined her spine.

"Hold still; this is going to hurt." I yanked it out quickly, and Birdie gasped. "Sorry. Thanks."

I used the feather to clean the mucus from Danea's gills, then checked on the guys. They were arguing with a couple of runty-looking orange beings with pugnacious faces. "Hey! Quit the small talk!"

Galena touched my arm. "Jory, did the slavers take our ship out of the well?"

"Since we're still breathing, I think that's a yes." I saw how frightened she was, and smiled. "They're going to wish they hadn't, sweetheart. I promise."

Danea opened her eyes and tried to push me away with a feeble hand. "Leave me. Leave . . ."

"Don't you give me any lip, Little Miss Mermaid, or I'll turn you into sushi." I checked her pupils and pushed the rapidly sprouting new growth of hair off her dark brow. Her terrified expression made me soften my tone. "We're working on getting some more water. Don't talk and try not to move around too much, okay?"

I took a few minutes to study our surroundings carefully. Something bugged me about this place, and not just the occupants. Then I began noticing things, little details that hadn't been apparent at first. My unease shifted into suspicion as I counted heads and genders.

The boys came back with a variety of split lips, bruised jaws, and limps. They also handed me enough shredded rags to cover most of Danea's feverish, shivering limbs. The last of the water barely dampened them.

However, what I'd spotted while waiting for the water convinced me everything was not hopeless.

"Have you seen any guards around?" I asked Kol, who was pacing a protective circle around us.

"No," he said, then turned as a scrawny, smiling prisoner scurried over toward us. "What do you want?"

"Heard you, heard you," the gaunt-faced slave chanted in singsong Jorenian. He had an unpleasant smile, one that revealed double rows of chipped, stained fangs. "No guards, no guards at all, at all."

"Someone must bring food and water," Osrea said, coming to stand beside Kol. Nalek took position on the other side.

"Not often, not often." Fang looked around before lowering his croon to a murmur. "Help her, help you, I can, I can."

Kol latched one clawed hand around the cringing slave's throat. "How?"

"Passage, passage, we go, we escape, there, there." He pointed his arm toward one corner of the cell. "After they sleep, they sleep." His arm swept out to encompass the other occupants of our charming accommodations.

"Why are you so hot to help us?" I asked, rising to my feet.

"Take me, take me, with you, with you."

His chanting began to get on my nerves. "What happens if we decide to stick around here?"

"Breeding pit, breeding pit." Three bulging eyes produced a lascivious gleam. "You breed, you breed, they sell, they sell, the young, the young."

Kol's face darkened and he lifted the slave off his feet, Nalek's fist clenched, and Osrea sputtered something obscene.

"Really." I scanned the cell once more, noticed the glint of a lens recessed into one wall, then nodded. "Thanks for the offer. We'll get back to you."

Once Kol placed him back down, the slave practically ran to the other side of the cell. All four boys took up immediate, defensive postures around me, Danea, and Galena.

"Kol." I got to my feet and pulled him to one side. "We need to talk."

* * *

Several hours later, after I told the rest of the clan what I suspected, and pointed a few things out to them, everyone went beddy-bye and our new friend Fang crept over to our little corner of paradise.

"Come now, come now, we go, we go."

Nalek picked up Sparky, while the rest of us carefully climbed over the slumbering bodies of our cell mates. The scrawny slave led us to the opposite corner, and gingerly rolled one of the orange-skinned midgets over before pressing a spot on the wall. A concealed panel slid to one side, revealing a dark passage behind the stone.

"Here, here, we go, we go."

We followed Fang into the tunnel, and waited as he closed the panel by pressing a spot inside the stone. He gave us a disagreeable grin, then skipped ahead, capering and chortling. The passage branched off in two directions, and he disappeared around the right corner. Almost at once he ran back when he realized we weren't following him.

"Why you, why you, not come, not come?" he asked, gesturing with his arms. "Hurry, hurry."

"We just have a couple more questions, pal." I nodded to the boys, who grabbed Fang and held him in place while I got up close and personal. "You said that cell we were in was a breeding pit. If that's true, how come Danea, Galena, and I were the only females in it?"

His three eyes practically popped out of their recessed sockets. "All dead, all dead, the other, the other, females, females."

I pretended to ponder this. "So what killed them?"

He shuffled his weight from one foot to the other. "Don't know, don't know."

"That's a shame. And these?" I tapped a finger against the implant in my chest. "Why don't any of the rest of you have these little beauty marks? Why just us?"

Fang gulped. "Don't know, don't know, you come, you come."

"No, I don't think we're going to do that." I patted his filthy cheek. "But thanks for showing us how to get out of the cell."

Kol's fist sent Fang to dreamland, and the boys carefully shoved the unconscious slave back into the cell.

"You must be right, Jory." Kol closed the panel and turned to me. "We should avoid whatever lies down the right passage. Let us make haste now."

We entered the passage opposite the one Fang would have taken us through, and began walking. The stone tunnel twisted and turned every ten meters or so, and complete darkness made navigating it a bit unnerving. Sparky was too weak to give off a glow, and the rattling sounds from her chest had started up again. Nalek refused to let anyone take a turn carrying her.

"First priority, find water," I whispered as the corridor seemed to widen, then took a sharp turn to the right. "Wait." Everyone stopped. I concentrated. "Did you hear that?"

Kol listened, too. "A mechanical plant? Drone units?"

The faint chittering reminded me of the insects and rodents who'd infested the underground tunnels. I'd hated listening to them almost as much as the confinement itself.

"Go slow now," I said. "And stay together."

The corridor's dimensions continued to expand, until it opened out into a large, empty cave. Warm, soft air wafted in our faces. At the other end I spotted a standard door panel set into the rock. There was no place else to go. I dropped my gaze, and found out why.

"Perhaps it leads to the surface," Osrea said, and would have stepped out into the cave if not for a quick grab by me. "What is it, Jory? I see no one."

I pointed down.

What appeared to be solid stone was actually an uneven lattice of thin strip-rock over an open, active pit of bubbling molten magma. The liquid wasn't high enough to spurt through the stone grid, but the threat of falling into it was obvious enough.

"Jory, what if you were wrong?" Galena peered at the magma and shuddered. "This could be the way the slavers keep prisoners from escaping."

"Oh, come on, Birdie. Modern slavers employ a bit more sophisticated means to keep their merchandise safe. You saw the recording drones I pointed out to you in the cell. Plus everyone in there was a little too filthy, scarred, and ready to kill. They don't look or act like slaves. They were bogeymen, put there to scare us."

"Perhaps they have not had time to adjust."

"They made it look like they've been there for years. Think about it—no slaver would waste profits that way. They sell whoever they capture as fast as possible."

Osrea eyed me. "You know a great deal about slavers, Clan-Sister."

"You should meet my former coach sometime." Let him think whatever he wanted. "I know that whole cell scenario was just what it looked like—a setup to make us *think* we were being held by slavers."

"But why?"

"I don't know," I had to admit. "But the answer is probably on the other side of that door panel."

Kol studied the obstacle. "We will have to do this in pairs. Osrea, take Galena. Renor, help Nalek with Danea. Jory and I will go first."

Hands linked, Kol and I stepped onto the widest piece of strip-rock, and discovered the surface of the rock was almost too hot to walk on.

"This is going to be like strolling through hot coals," I said, lifting one foot, then the other to keep them from staying in contact with the strip-rock too long.

"Do not try to run." Kol carefully tested his weight as he walked a few steps forward. "Come, it will hold."

And it did, for a few more feet. Halfway over the pit, the strip we stood on started to vibrate under us. I glanced back to see the others already too far from the edge to jump back.

"Now would be a good time, Kol," I said as some of the outer strips crumbled, "to *run*."

I didn't run as much as let myself be dragged across the remainder of the strip-rock. Kol left me at the door panel to reach out to Osrea and help him with Galena, who was already in his arms. Nalek lost his balance and literally tossed Danea to Renor before falling down between the strips and grabbing the edge of one to keep from dropping into the magma.

I yelled. Kol reached for Renor, helped him carry Danea the last couple of feet, then leaped back out on the lattice. Nalek was holding on to the rock with one arm, and swung the other up to reach for Kol's extended hand. It took a minute, but Kol tugged Nalek back onto the strip. Just in time, too. That strip began collapsing, and they were forced to jump more than a meter to reach the edge.

I didn't start breathing again until they made it. By that time the entire lattice had collapsed into the pit of magma.

"Jesus." I sat down beside Danea and cradled my aching knee. The bottoms of both my feet had mild burns, too, but I didn't care. Nalek was safe. We were all safe. "If this was the safe tunnel out, I'd hate to see what Fang would have made us go through."

"We have to keep moving." Kol got up and headed for the door panel. "There are no access controls on this side."

"Let me try." Nalek had some minor burns on his legs, but that didn't stop him from pitting his considerable strength against the plasteel panel. It began to slowly inch to one side. With a groan, Nalek forced an arm through the narrow gap he'd created, then shoved the door wider until it slid completely open. Whatever he saw on the other side made him step back.

"What is it?" I moved over so I could see through the opening. A figure shrouded in black stood waiting just inside the panel—not the M.O. of any slaver I'd ever heard of. "Shall we introduce ourselves, guys?"

Kol and Nalek rushed through the panel, followed by Osrea. There were immediate grunts, groans, and thuds. I motioned Galena and Renor, who was carrying Danea, back, then crouched over and rushed in.

Nearly everyone was on the ground and holding some significant portion of their anatomy, with the black-shrouded figure standing in the very center. I came to a skidding halt and took a defensive stance, but I didn't see any weapons. Evidently he'd knocked all three of them out.

"Who are you?" I shouted.

"Jory. Let us not battle here." Renor walked in and inclined his head briefly, then said to the black-shrouded figure, "One of our females requires immediate medical attention."

As the boys got to their feet, our captor made a shallow bow, then walked into one of the entrances without uttering a single word. Other entrances to several corridors branched off in different directions.

Now I was totally confused.

"Do we follow him?" I asked Kol, who appeared to be seething.

Muscles started bulging as he flexed his arms. "I plan to."

We went into the entrance after the figure, and walked a short distance into what looked like a formal reception hall. Sparse decorations made odd patterns on the walls, but there were no furnishings and only an inset com unit next to a door panel. The black-shrouded figure stood there waiting until we all appeared, then tapped something on the panel, which opened.

Another figure in black emerged.

"I see you have escaped from the breeding pit," a low, arid voice said from beneath the featureless mask. Masculine, from the sound of it.

"If that was a breeding pit, I'm a swarm-snake," I said. "You the head flesh peddler?"

The mask tilted to one side. "I beg your pardon?"

Kol stepped up to eye the new arrival. "Where are we? Who are you?"

"I am Dursano, second-level inductor." He lifted a glove and removed his mask to reveal an austere, humanoid face. "I will be briefing this group on the next level of your training."

Osrea stepped in front of Galena. "You're not touching our females, slaver."

"We need to find an infirmary, Os," I said, then gestured toward Danea as I addressed Dursano. "Our ClanSister needs medical attention."

He nodded. "I will see to it at once. Here." He produced a stack of black garments. "Put these on."

No one made a move to take them. We'd become a suspicious bunch after what had happened in the slave pit.

"What do you mean, next level of training?" I looked at the bare chamber. Maybe this was some sort of test to determine how resourceful we were. "What's this second level?"

"You have done extremely well thus far. Not many students have successfully graduated from the first level within the first week of training."

Whatever answer I was expecting, it wasn't that. Students, training—what the devil was he talking about? "Where are we?"

He went to a wall console and pulled up a split image on the vid screen—one of our prison cell, and another of the magma pit. He touched the keypad, and both vanished.

In their place were bare rooms with odd-looking grids on the floors and walls.

"They are dimensional simulators," Kol said.

"That is correct."

I rubbed the back of my neck. "So you're not slavers, and this isn't Garnot."

"You are on Reytalon." He turned and inclined his head. "Welcome to the Tåna."

CHAPTER ELEVEN

"HONORING THE PATH IS AS MUCH DUTY AS BURDEN."

—TAREK VARENA, CLANJOREN

We got dressed in the formfitting garments, which were black, but not made of the N-jui dimsilk that the dancer on the *Chraeser* had worn. This fabric felt a lot like Terran cotton. The clothes also left our hands, feet, and heads bare. Once we were all decent, Dursano led us down a corridor to a large, well-equipped medical unit—a completely empty unit, except for two nurses and a physician, who immediately tried to take Danea from Renor.

I stepped in. "She's an aquatic; she needs water." A glance around didn't reveal any large amount handy. "Where are your immersion tanks?"

"Give the female to us." The doctor, a flat-faced alien with an abbreviated horn in the center of his square brow, glared at me. "We will deal with her treatment."

Nalek stepped in front of Renor and folded his big arms, making the sleeves of his shirt stretch. "You will show us the immersion tanks."

Horn Head looked at Dursano, who gave him a slight nod. Then the physician made a choppy gesture toward the back of the facility. "This way."

Nalek went with Renor. Kol watched them go, then addressed Dursano. "You owe us much in the way of explanation, Inductor."

"Yes." Dursano waited until Nal and Ren had rejoined us. "There is a briefing room across the corridor. Your comrade will be safe here."

"Ren," Kol said. "Stay with Danea."

The inductor made no objection, and led the five of us into an anonymous meeting area across the hall. Despite the empty chairs around a large dimensional projector, no one sat.

Dursano took a position at the projector control panel. "The aquatic requires time for proper treatment. When she is recovered, your group may enter the second level of training."

"Really. After being captured, operated on, and thrown in prison?" I didn't bother to hide my sarcasm. "If that's level one, I can hardly wait to see what's next."

That got me a bland look. "You contracted transport to Reytalon, did you not?"

"Yeah, but our contractor changed his mind and sold us to slavers." I paused. "Or was that part of the training setup, too?"

"For security reasons, I cannot discuss our induction methods," the inductor told me. "However, if any of you wish to leave Reytalon, you may do so at this time."

"We are not leaving." Kol moved toward the console. "Not until you explain what you have done to us." He touched the place on his shirt that covered the new scar we all had. "Begin with these implants."

Dursano's lean, four-fingered hand touched the panel, and a holoimage of a round, flat object appeared above the console. "Each of you has been given a proximity implant, placed over your primary cardiac organ. The device provides tracking information on all students. It also records training bout 'kills.' "

"Asking if we wanted them would have been nice," I pointed out.

"We do not make requests of those who come to Reytalon. You endure, you leave, or you die." Dursano smiled briefly at our collective reaction. "Perhaps it is better that I begin with the nature of your training."

He waved his hand over the panel, and five apertures opened, each producing what looked like a bladeless hilt.

"These are tåns, holographite transmutational weapons used to train our students. Take them."

Kol picked one up and examined it. So did I. The material of the hilt felt odd. I realized why when I clutched it in my fist and it adjusted to my grip. Biomalleable hilts. Just what every trendy assassin needed.

The image of the implant disappeared, replaced by a projection of a tån, this time with a long, heavy blade attached to it.

"Students use the tån to train and learn combat skills through the *shahada*. The weapon goes through seven stages of transition, beginning with raen-tån, or great sword."

"How do you change the size of the blade?" I asked.

"Transmutation controls, located on the pommel." Dursano pointed to tiny, recessed spots on the curved butt of the hilt, then added several other images of different-sized blades. "As training levels are achieved, our students learn to wield progressively shorter blades, seen here. The thion-tån, rangi-tån, jyan-tån, shou-tån, and elok-tån." He erased all but the smallest image, which abruptly divided into two. "In the final stages of training, the student masters the osu-tån, or dancing blades."

"One for each hand," Kol said.

Dursano gave him a sharp look. "Yes. The discipline of the tån ultimately leads to close-proximity, two-handed fighting. Using the osu-tån is the deadliest of all dances."

"Yeah, I can see me running into hordes of two-handed knife fighters," I said, disgusted, and tossed the tån back on the console. "So this is it? Just fancy blade tricks? You don't teach more advanced skills; how to use other weapons?"

"There are none. There is only the blade."

"Oh, come on." I gestured toward the tåns. "A trained soldier with a pulse rifle can take down any idiot armed with one of these, given enough distance."

"Indeed." Dursano erased the image of the two blades, and bent down. A moment later I reflexively caught the displacer rifle he'd thrown at me. The instructor walked out from behind the console and made himself a target by standing a few feet from me and spreading his arms. "Prove your words, Sajora Raska. Shoot me."

I laughed. "I don't think so."

"Then you are a coward who will fail in all things."

I didn't like being called a coward, so I lifted the rifle and aimed for his upper right arm. As I pulled the trigger, Kol yelled, and Dursano moved slightly.

Something sliced across the back of my hand and knocked the rifle to the floor. The sound of something thunked behind my head, and out of the corner of my eye I saw a knife imbedded in the wall panel. I grabbed my wrist and looked back in time to see the instructor cross the distance between us with some weird, fluid, rolling movement. A heartbeat later I was staring down the very short length of another blade he held to my carotid artery.

The entire thing had taken maybe three seconds.

Dursano pulled the osu-tån out of the wall. "You were not fast enough to kill me, Sajora Raska."

No, I wasn't, but that wasn't what bugged me the most. "That blade should have sliced my fingers off. Why didn't it?"

"Holographite blades dematerialize upon contact with any student's body."

"Not instantly, or it wouldn't have cut me."

"You are bleeding," I heard Kol say in a low, dangerous voice.

"Do not attempt to attack me, Jakol Varena." Dursano never moved as Kol came up behind him. "The blade I hold at Sajora's throat is *not* holographite."

"It's okay, Kol. No harm done." I flexed my hand. "I'll need a little bandage, that's all." I regarded Dursano. "How did you do that move? What's it called?"

"I am a blade dancer. The move is called the Banner Extends." The instructor removed the knife from my throat. "If you choose to continue, you will learn thousands of moves like it in the *shahada*."

"And what do you get out of it?"

"Should you graduate, we offer employment, and membership in the order." He sheathed his knife. "Whether you serve the Tåna or not always remains your choice."

Dursano didn't press us for an answer, but sent us back to the medical unit. The horn-headed doctor cleaned up my wound while the others went to check on Sparky.

"Well?" I asked when Ren appeared with the others—minus Nalek. "Is the disagreeable shrew all right, or what?"

"Danea is conscious and appears to be recovering." Ren's crystalline face glittered as he grinned. "Her first words were, 'Tell that deranged Terran female I am fine, no thanks to her.' "

I laughed. "Sounds like she'll live."

After we were served a plain but substantial meal, Dursano reappeared and took us to yet another chamber for the remainder of our briefing. The instructor reviewed the main points of entering second-level training, but revealed nothing about what lay ahead of that.

"Students are ranked by color bands. White is the lowest rank, which you will wear." He tossed a handful of white bands on the console. "Followed by yellow, orange, red, green, blue, and purple."

I picked one up. "Rainbow outerwear. How cute."

"Placement is symbolic and a personal choice. Around the thigh, soldier for hire; the arm, defender; the brow, warrior. Perhaps you seven will wear yours around your necks."

"What does that mean?" Osrea asked.

"Assassins."

We also learned our sleeping quarters, meal intervals, and cleansing units were communal, and that we would be allowed only four hours of sleep each rotation. The school's curriculum was simple: a blade dancer killed. We would be taught to use the tån in all its forms for that purpose, and that purpose only.

"How do we advance to the next level in training?"

"To attain the next color band, one must defeat two next-level students in training bouts. This is for all ranks except purple and black. To attain purple, one undergoes the physical trial run."

"What about the black band?" Galena asked.

"The order of the black is worn only by Tåna graduates, who have entered the quad during the Tåna-Shen and defeated twelve white, ten yellow, eight orange, six red, four green, two blue, and one purple."

Kol's white eyes narrowed. "All of those in one bout?"

"Of course. That is the only way to join the order."

We all had questions—who wouldn't?—but Dursano declared the session over and sent us to our temporary quarters at the end of the corridor.

"Enjoy these last hours of comfort," he said as we left. "You have until the time the aquatic female has recovered to choose to leave Reytalon. Should you decide to remain, you will join the others in the second level."

"How long does it take to go through all the levels?" I had to ask.

"It takes until you are ready to graduate and join the order, Sajora Raska." Dursano pulled the feature-obliterating mask over his face. "Or until you die."

Our hours of comfort lasted exactly two rotations. We spent them taking turns watching over Danea in the infirmary, eating, sleeping, and otherwise collecting ourselves for what lay ahead.

Oh, and we argued, too. That was becoming a given.

"We got this far, Snake Boy," I said as I checked my knee brackets. One of the nurses had grudgingly given me a caliper that made adjustments a breeze. "I'm not backing out now."

"He said *until we die*," Osrea said, and ripped two more holes in the side seams of his student tunic. With a sigh of relief he eased his extra limbs through the tears. "I came here to learn to kill. Not *be* killed, ClanSister."

"Nor am I enamored of the thought of killing other students," Ren put in.

"He probably said *die* meaning of old age." I pulled down my thermals and checked the room console. "Time to relieve Nalek. Whose turn is it to baby-sit Sparky?"

"Mine." Galena, who never liked it when we fought, jumped to her feet and headed for the door panel.

I wasn't letting her off that easy. "Birdie, do you want to stay or go?"

She skidded to a halt and her winglets arched. "I will stay." Out she went, as though chased by a herd of stampeding *t'lerue*.

I eyed Osrea. "Guess that means *you're* staying, right, pal?"

"I am not your pal." Snake Boy stomped off to the prep unit and dialed up some disgustingly raw concoction.

"They will attempt to separate us, will they not?" Renor asked no one in particular.

Kol ran a hand over the dark stubble that had appeared on his scalp—Jorenian hair grew back fast. "Perhaps. We will not allow it."

"How much we will be allowed to do remains to be discovered." Ren went to the viewport and stared at the static white blizzard raging just beyond.

Nalek entered, looking happy. "Danea has regained her mobility and her corporeal field is fully recharged. She will be released in the morning. What say you, ClanSiblings?"

Ren went back to staring at the surface. Osrea grunted something over a mouthful of uncooked food. Kol merely nodded.

When Nalek turned his disappointed gaze toward me, I shrugged. "Sorry, big guy. Not everyone is thrilled about the idea of going into training tomorrow."

"Danea is. I left her still arguing with the healer about discharging her so she could continue on." Nalek seemed mystified by our group depression. "Have we not met every challenge with courage thus far? Are we not united in our convictions?"

I went over and patted his shoulder. "We're working on it; we're working on it."

Os put down his server of uncooked glop. "I will enter training willingly if you convince Galena to return to Joren."

Kol eyed me. "Sajora can return with her."

"Danea will be hard to convince, but perhaps I can persuade her to accompany Galena and Jory as their protector for the journey home," Nalek said, his expression fading from exuberant to thoughtful.

I nearly choked on my tea. "Excuse me? You want us to go back because we're *female?*" Kol and Nalek gave each other sideways looks. "Oh, give me a break. Like anyone on Joren is going to welcome us back with open arms. They probably threw one of their ClanParties as soon as we left." I studied their expressions. "You're serious about this. You think because we're women we can't make it through this training? What happened to the famous nonexistent gender bias of the wonderful Jorenian people?"

Ren offered the logical arguments, naturally. "Galena's physical strength is limited. You have a substantial disability. Danea remains weak from her ordeal."

"Feeble little Galena was the one who got us out of Uzlac's prison chamber, remember? Danea doesn't even need blades—she can jolt the daylights out of anyone who gets in her face. As for my *disability,* it didn't prevent me from beating Kol in a warrior's quad on Joren."

Kol scowled. "You tricked me."

"Still worked, didn't it?" I planted my hands on my hips. "All right, cut it out, you guys. We're in this together. Either we all go into training, or we all leave. Since I'm the only female here, I'll vote for all three of us. Birdie, Sparky and I are staying."

"I will stay," Nalek said.

Osrea grumbled something that sounded like a "Me too."

Renor didn't turn around. "You are my HouseClan. I stand with you."

Kol gave me the once-over, then nodded.

"Great." How many more times were we going to do this? I figured on a couple of million. "Let's get some sleep while we still can."

A day later we stumbled into our newly assigned quarters, more dead than alive. None of us stopped moving until Sparky tested the floor. One of the first things we had learned about the Tåna was that they kept you moving—or else.

"They spoke the truth. This floor is not charged," she said in a low, rasping voice.

"Good." Since there were no berths, only a couple of thin mats on the floor, I fell on the first one I got to with a thump. My bruised, abused body screeched in protest, but I was too exhausted to care what damage I'd done. "See you in four hours."

I couldn't sleep at first—I hurt *that* much—and I had a lot to think about. Like everything that had happened since we'd left the infirmary earlier that day.

Dursano had collected us at our quarters, distributed our holographite blades, and instructed us to carry them in osu-tån form at all times. Evidently it was some kind of dancer tradition—if you lost one blade in a fight, you'd still have a backup. The entrance to the second level of the Tåna, at the end of the corridor outside the infirmary, opened into yet another a long, featureless hall.

Before we entered, he held up one hand.

"Blade dancers know two fundamental truths. The first is that they will die. The second is to live each moment as if death awaits them in the next." He looked at each of us before adding, "Do not enter unless you are willing to learn this, and embrace it."

For a welcoming oration, that was pretty profound. So much so that it rendered us all speechless, which was the point, I suppose.

Then came the beginning of all the fun stuff.

Nalek was the first to discover we couldn't stand in one place for longer than five seconds. He did that by dropping the white student band he was carrying, and bending to pick it up. Then he yelped and jumped. I'd never seen a guy his size jump that high.

We all stopped. Dursano and the instructor did not.

"The floor!" He stared at it, aghast. "It shocked me!" Then he jumped again, and so did the rest of us, as a biocharge jolted up through the soles of our feet.

"Keep moving," Dursano called back without looking.

I caught up with him. "This place has electrified floors?"

"Yes. Contact with any common area floor surface outside your student quarters for more than five seconds produces a minor shock."

"What is a common area?" Kol demanded.

Dursano made an encompassing gesture. "All of the corridors, and any area where students might otherwise loiter. All common areas are clearly marked." He pointed to a thin red line I hadn't noticed before on both sides of the floor.

"How nice." I snorted. "And what happens after the first jolt?"

"Continued contact causes more serious charges, until the offending student is rendered unconscious and demoted to the next lower level of training."

In our case, that was the dimensional prison cell. Or did they do something worse when you flunked out? "You consider people standing in one place a bad thing?"

"For a blade dancer, it can be fatal."

Our tour of the Tåna's second level went on. The empty connecting corridor opened out into a wide training floor filled with students dressed in black garments identical to our own. Unlike dimsilk, the clothes clearly displayed body outlines, and underscored how many different species there were—everything from humanoid to insectile, reptilian to aquatic.

"This reminds me of the zoological exhibit in Lno," Osrea muttered.

I smiled. "They probably think the same thing about our little group."

Some of the students were sparring in pairs, but most were gathered around an elevated quad, where two figures were fighting in earnest. No red lines on the floor here, so watching matches obviously wasn't considered loitering.

"Level two training begins here, with instruction in movement." Dursano indicated the first of seven door panels lining the walls around the main floor. "You will also receive instruction in targeting, grappling, timing, stealth, and escape. Once you have mastered these basic techniques, you will begin bladework"—he pointed to the last, center panel—"and fight other students in quad training. I will introduce you to your trainers now. Come."

Dursano led us to the first door panel and opened it. Inside, a metallic drone occupied the center of the room, while a group of students moved around it. Feeling a bit paranoid, I checked out the floor, but saw no red lines.

The drone had been mounted on a revolving platform, and

its many discharge ports were shooting out thin pulse beams at random angles. The students were apparently trying to dodge the beams.

Dursano stepped in and called out, "Halt exercise."

The drone's ports stopped firing, and the students formed a perfect circle around the platform.

The drone dismounted the platform and folded its elongated limbs into a more compact shape. "How may I assist you, Inductor?"

"I bring new additions for the roster." Dursano named each of us, and had us speak to the drone so our tone patterns could be recorded for future reference. "They will voiceprint for the other trainers, then return to begin."

"Acknowledged."

When we left the class, Danea stepped in front of Dursano. "Why must we train with drones? Have you no *real* instructors here?"

"You are only worthy of drone trainers now," he said. "If and when you assimilate the proper skills worthy of advancement, you will move to the final level of training, where you will be trained by members of the order."

We moved on to the next class, which was unoccupied, and repeated the process. The drone trainer in this room was similar in design to the previous one, but had shorter limbs and wheeled around the floor instead of occupying a central platform. The third trainer had a complex array of artificial limbs and appendages, all geared toward demonstrating a wide variety of grappling techniques, while the timing trainer was a tiny, incredibly fast hoverdrone that darted through the air.

The last two classes—stealth and escape—were held in automated dimensional simulators. For these, we each submitted to a retinal print and the programming panel.

"Your trainers will direct you to the appropriate class, communal areas for meal intervals, and dismiss you to your quarters at the end of each training rotation. You will not deviate from this schedule." Dursano led us back to the movement trainer's room. "Enter and begin."

* * *

From there things went straight to hell.

In the movement class, we were grouped with another ten students and instructed to gather around the center platform in a circle. The drone trainer gave us a brief set of instructions.

"In this period, you will be taught how to move, to counter attacks, and to avoid injury. My portal beams are programmed to deliver neural shocks, impact contusions, or dermal cell burns. Intensity and duration increase with successive hits. Your sole objective is to avoid the beams." The drone mounted the platform and settled itself into place. "The session begins."

The first series of beams shot out of the portals, but they were widely spaced and simple to avoid. I began to wonder if this was really worth my time when the drone spoke again.

"When encountering your target, assume they are armed with an energy weapon." A fierce volley of concentrated beams erupted, like weapon fire, in clustered bursts. None of us were hit, but only by sheer luck. "Assume others in the vicinity are also armed, and sympathetic to your target."

I heard something hiss behind me, and whirled in time to see a wall panel open and more portals appear. "Kol!"

He whipped his head around, then swept a hand toward the floor. "Down."

All of us dropped to the floor except Osrea, who crouched over and yelped as one of the wall portal beams struck him in the back. "Mother, what—"

"Move." Kol rolled to his feet and pushed him forward.

Beams were shooting everywhere by then, and I grabbed Kol's arm to pull him out of the path of one.

"Pair up," I said to the others. "One watches the walls, the other the drone."

We managed to avoid most of the beams that way, but nearly all of us were hit once before the session was over. When the last of the beams faded, the drone dismounted the platform, directed us to stop and form a circle, and then divided us into three groups. All of us except Renor.

"Had this been an actual encounter, you," it said, indicating the largest group, which included Galena and Osrea, "would all be dead. You"—it turned to the second, which Nalek and Danea had been placed in—"would be seriously injured and unable to escape. You," it said to me, Kol, and two other students, "would have minor injuries, which may leave trace DNA evidence at the scene."

The drone then went to Renor. "According to the attack grid, you should be dead. However, I read no indication of any beams making contact with you." It paced around him. "Had your derma reflected the beams, the feedback would register, and they would still count as hits. I am unable to classify or rate your performance."

Renor only stood there and said nothing.

"You are dismissed to targeting instruction."

As we walked out, I noticed Danea giving Renor a hard look, and wondered just how powerful Plas-Face's telekinetic gift was.

Targeting class made dodging beams look like a walk in the park. We were each issued white, tunic-styled overgarments by the trainer and told to put them on. The problem was, there were no sleeves in them, and the effect was like shrugging into a straitjacket.

"These are binding shirts. In this session, you will not be permitted to use your arms."

"Why not?" asked Os, who looked like an octopus with the ends of his multiple limbs sticking out of the shirt.

"You will learn how to target by becoming targets yourselves." The trainer turned to admit a group of other students, all carrying their tåns in the raen, or long sword, form. "These students are attackers, and will pair off with each of you. Your objective is simple—avoid their weapons."

I had almost fallen asleep when I sensed a small form hovering over me. It was Galena, holding one of her winglets.

"Jory?" She knelt down and turned a little. "Your pardon, I did not wish to disturb you, but I feel as if I am torn here. See you anything that is bleeding?"

I rolled onto my side and peered at her back. The thin cord of tissue along the top edge of her wing was twitching and knotting.

"Cramps," I said, and sat up to deal with it. I rubbed firmly, ignoring her gasps as I worked the strained muscles. "You did pretty good in targeting today."

"You mean I can run fast." She sounded wry. "How is your cheek?"

"It hurts." I'd gotten a sizable bruise during grappling class, when the drone trainer and I had gone one-on-one. I was more concerned with the abnormal way she was holding her winglets. "Birdie, are you doing this on purpose? Bending them in like this?"

"It is how I have always held them."

"No wonder you're cramping. Here." I gently stretched out the ligaments, but she resisted. "Stop trying to flatten them against your back. Let them arch naturally."

She did, and sighed. "You are right; that is better."

"I don't know why you're still trying to fold yourself up like that." I looked around at our exhausted companions. "Everyone already knows about your wings, sweetheart."

"I saw others like me—avatars—today." She bit her bottom lip. "There were a whole group of them flying in the simulator classes."

I thought of the mud swamp we'd trained in. "You could see in all that muck they made us wade through?"

"I cannot help watching the skyline."

Why would she be worried about other bird-people? "Maybe they were part of the program."

She shook her head. "They were real. I saw a pair of them walk out. They looked at me." She swallowed. "They will ridicule me, will they not? When they find out I do not know how to fly?"

We'd all gotten a hefty dose of ridicule and scorn from the other, more experienced second-level trainees. Most of them stared and laughed at us as we stumbled through the classes, but a few had made aggressive moves against us, particularly us females. As we went from one class to another, they pretended to accidentally bump into us, brushing against us from behind and muttering suggestive propositions. Birdie had pretended not to notice, but I'd shoved them away and told them out loud what they could do with their various sexual organs—or what I could, if they kept it up.

Danea hadn't been obliged to do anything. A single contact was enough for everyone who tried to mess with her.

"Galena." I nudged her chin up. "You've got some feathers to grow back before you try out these wings, but who's to say what kind of flier you'll be? Maybe you'll be faster than they are. Don't let them intimidate you."

"I wish I were larger and stronger. Like you." She smiled a little. "They were scared of you and Danea and Kol. Especially Kol."

"Hey, *I* was scared of Kol." And I had been, when I'd seen what he'd done to the first idiot stupid enough to try to capture him with a net in the simulated tunnels of the escape class. The student, who hadn't possessed wings, had gone flying along with his net through the simulated tunnel wall and had to be hauled off to medical for some stitching. Everyone had stayed well away from Kol after that. I felt the tension in her back ease away, and stopped rubbing. "Better now?"

"Yes. Thank you, ClanSister."

I patted her thin cheek. "Get some sleep, little bird."

Galena returned to her mat, stretched out with a sigh, and quickly fell asleep. Over her shoulder I saw a pair of glowing white eyes watching me.

My mouth hitched. "You want to share some girl talk now?"

Danea rolled over and presented her back. "Shut up, Terran."

"Good night, Sparky."

The second day of Tåna training was only slightly worse than the first, objective-wise, but we were all tired, and that made us prone to making more mistakes. Even Renor, who didn't sleep and hardly messed up the day before, seemed to be running on empty cells.

"What's the matter with you?" I asked him as we dodged more raen-tån-wielding students in targeting class. He'd been "killed" at the end of movement class, and so far had taken three hits as a target. The latter didn't harm any of us—the holographite blades dissolved on contact—but one of the students had struck dangerously close to Ren's chest implant.

"I have overextended my . . . self," he said, the planes of his face beading with sweat.

"So has Sparky, apparently." Danea's corporeal field seemed to have lost its sting, and I'd noticed her taking more of a beating today as well. "You guys need to conserve your zings a little more."

Renor flashed like a Yuletide ornament as he spun out of harm's way. "That would be wise."

I waited until my attacker was nearly on top of me, then flipped out of the way and sent him sprawling with a well-placed kick. "You'd better tell Sparky to conserve her batteries, then, 'cause she won't listen to me."

Kol stopped us from heading in the same direction as the other students when we were dismissed from grappling, and pointed to the long way around the quad. With a few raised brows, we followed him.

"I overheard two of the others talking about a female named Fayne," he said. "They said she will be confronting one of us today."

"Everyone in this place is confrontational," Nalek pointed out in a reasonable way. "Why should this female concern us?"

"Evidently she has a reputation for brutalizing new arrivals and dispatching them from training." Since we were in a red-lined area, we couldn't stop moving, and Kol led us in a second, unnecessary lap around the quad. "Stay in pairs—Nalek with Galena, Osrea and Danea, Renor with Sajora. I will take point."

"Excuse me, fearless leader." I tapped him on the arm. "What if she goes after you?"

"I will have the six of you at my back. I am hopeful you will watch it." He hissed as a beam connected with the top of his shoulder, and a hoverdrone appeared. "It appears we have deviated long enough from our training schedule."

The drone trainer was already lecturing the class on the techniques to be used that day, but tracked us with his optic sensors as we assumed our positions among the ranks.

"—never delay locking a grip on your opponent, but establish the one-handed hold as early as possible." Lights flickered along its metallic abdomen to show contact points as it explained the next move. "If your torsos are touching, wedge your lower limb joints between any he may have, bend as if to prostrate yourself, then push your left limb out quickly and forcefully, and fall upon him." The drone's head turned. "We will begin practicing the various grip-falls with those who were unwilling to arrive when this session initiated."

That meant us, and the other students made various sounds of amusement as we approached the drone.

"Nal, Os, Kol, Ren," the drone recited, shortening their names to one syllable as the other trainers had. "Approach me from the front. Dan, Gal, and Saj, from the rear. Establish a locking grip and attempt to pin one of my limbs to the floor."

A minute later, the trainer had all seven of us wedged in various positions on the floor under his web of jointed appendages, and the class erupted into outright laughter.

"I trust you will be on time for the next session?" the trainer asked as it released its many grips and let us up. "Yuz, San, Hil, as you find their efforts so entertaining, you may now attempt the same."

Knowing someone was gunning for us made us more alert as a group, and we stayed paired up and watchful for the remainder of the training day. Only after we left the escape class simulator to head for our meal and rest interval did I start to relax.

"So much for the big, bad Fayne."

"It could be that she had second thoughts," Galena said as we passed by the more experienced students emptying out of the bladework room.

"Or perhaps she waits in the communal eating facility, hoping to ambush us," Danea said, her corporeal field as dismal as her tone.

Nalek and I were flanking Kol from behind, but I decided a practical example wouldn't hurt.

"We need to establish a rep, that's all." A short, blocky humanoid with an odd, pointed skull got in my way, then turned in toward my left side—a definite sign of aggression. "Like right now."

I rolled in and under its guard, using my hip along the way to knock it off balance. It lashed out at me with a powerful upper limb, but I grabbed a handful of garment and used the momentum to send it into a spin. The tunic tore, and the humanoid went staggering into a wall panel. It hit with a very satisfactory thunk and slid to the floor.

I regarded the swatch of black material in my hand before letting it drop. "You ought to watch your step, Pinhead."

Other students gravitated toward us, but were thrust aside by a platinum-haired humanoid wearing a green band around her neck, followed by a shiny black insectile being and two behemoths covered with long, dirty-looking gray fur. All four headed straight for me, and they didn't look happy.

"Kol." I kept my voice mild. "Clumsy here has friends."

My clan gathered around me, keeping their backs in so they could face the gathering crowd on all sides. Before Pinhead could scramble to its feet, Blondie grabbed him by the arm and heaved him up. The two walking rugs took hold of him as she faced me.

I considered the potential threat the green presented. Her pretty, dramatic face was surrounded by enough hair to show she'd been at the Tåna for a while. Not much of a forehead, but the back of her skull swelled out in two distinct curves. Her eyes were as black and polished as her bug friend's shell, and looked bottomless compared to her dead-white skin. The top of her head barely cleared my waistline, and yet the little female had her tåns in hand and stood poised to strike.

My blades were already out, but I gave diplomacy a shot first. "You have a problem?"

"You are breathing," she said, in a surprising voice pitched even deeper than Nalek's.

So much for the tactful approach. Pinhead and Blondie had

plenty of friends, from the number of angry-looking students who began popping up around the behemoths, forming a wall to keep anyone from walking away.

Kol and Nalek closed ranks on either side of me, and I glanced down for a moment to see their blades out and waiting. The girls closed in on our sides, while Ren and Os watched the rear. We couldn't remain still, and moved in a slow clockwise pattern. Blondie, Pinhead, and their wall of pals likewise mirrored our movements.

"I want the Terran," Pinhead said, panting.

"No." The green flicked her blades, a razzle-dazzle trick that made them spin like wheels. "She is mine, Cirilo." She took a step toward me.

A hoverdrone appeared between us, just above Blondie's head. "Has a formal challenge been offered and accepted?"

"Not yet," I told it. "I don't even know her name."

"That's the current Tåna quad champion," one of the students to the side said in a helpful way. "She is called Fayne."

"*You're* Fayne?" My laugh was spontaneous. "But you're just a midget."

No one else thought that was funny, and Blondie's glittery eyes constricted to thin slits. "Your tongue will be the first thing I cut off."

The drone dropped and swung up by Blondie's shoulder. "If this is a formal challenge, declare it."

I used the momentary distraction to feign sheathing my blades, at the same time swinging and driving my elbow into Fayne's face. She staggered back, bounced into one of the behemoths, then regained her footing.

"Whoops." I kept my tåns between us. "Clumsy me. What was that about my tongue again?"

"You." She wiped a thin trickle of clear fluid—maybe her kind of blood—from her lip, then held up a hand as her cronies shuffled forward, stopping them cold. "You dare to strike me." The drone inquired about a challenge again.

"Stand by," I told it. "She's still a little confused. The confusion may go, but the little's for life."

She blinked. "You know who I am."

"Sure, but the important issue is, do I care?" I let my gaze drift down, then back up. "Not really."

Cirilo lunged forward, only to be brought up short by Fayne as she snapped out an arm and smacked him in the chest.

"No. Not here." She gave Kol a long, interested look, then refocused on me. Her close-set black eyes had double eyelids, and one slowly dropped, then raised as her small mouth lifted at the corners. "This is what our Blade Master admits to training so that we may practice our craft. Clods and dregs. Dead carcasses present more challenge."

I laughed. "You get beaten up by dead people regularly? Maybe you should find another line of work."

Fayne moved forward then, but a crackle of energy flared as Danea came to stand shoulder-to-shoulder with me.

"A fish breed." Blondie licked her lips with a little white tongue. "We eat your kind on my world."

Danea turned her blade so it reflected a flicker of light. "Come and have a taste."

"Or you can bite me," I offered.

"Dregs always boast from behind their comrades." Fayne turned back to me. "And you, clod. I will enjoy cutting you. I will take my time."

"You will either issue a formal challenge or disperse at once," the hoverdrone demanded.

"I just don't see much of a challenge here." I nudged Danea, ignoring the brief jolt. "Sparky? You want to waste *your* time?"

"She has a large mouth." A tendril of her yellow hair crackled with energy, less than a centimeter from my cheek. "Nothing else impresses me."

"No one wants to play with you, Blondie." I flicked my fingers. "Run along now."

"Our time will come." Fayne gestured to her cronies, and they turned their backs on us en masse to follow her to the communal eating facility. Cirilo hesitated for a moment, then hurried after them.

It took another minute for the rest of the spectators to disperse, but not without a lot of troubled looks at us and muttering among themselves.

Slowly I sheathed my blade. "She's trouble."

"Hmph." Danea's hair settled down. "She is small."

"Regardless of her stature, she has many comrades." Kol looked thoughtful as he watched Fayne and her group disappear. "And there is something about her. Do any of you recognize her species?"

"Fayne is a Skogaq. Species data is available on your room console." The being who said that strolled by us, its form completely shrouded in dimsilk. As it did, it used the *goreu* staff it carried to point to the dining hall. "You have only ten minutes left for your meal interval. Go."

"And you are . . . ?" I asked.

"Uel. Blade Master of the Tåna." It continued on and disappeared into one of the corridors.

What do you know, I thought. *Just the man I need to talk to.*

CHAPTER TWELVE

"THE TRAVELER ENDURES THE JOURNEY; THE PATH ENDURES THE TRAVELER."

—TAREK VARENA, CLANJOREN

"Okay, she's fast," I said as I paced the perimeter of our quarters thirty minutes later. "Big deal, so am I."

"Skogaq are not merely gifted with speed." Renor finished reading the last of the data. "Her kind have unusual physical conformities—hollow bones, multiple-jointed limbs, extremely powerful tendons and ligaments—"

"So she's fast *and* stringy." I stopped when Kol got in my way. "What?"

"You will not challenge her." He swiveled to look at the rest of the clan. "None of you will. This female delights in killing."

"She's nasty, a midget, and a bully." I went around him. "Sounds like a good reason to pick a fight to me."

"She is the Tåna champion. You have not her skill level with the blade, Sajora. None of us do." Nalek stretched his arms over his head and yawned. "We should sleep."

"We should settle this first." Osrea countered my pacing with his own annoying habit of thrumming two sets of his fingers on the floor. "You embarrassed her, Jory, in front of those who fear and respect her. She will take revenge, if only to preserve her status among them."

"I agree. Her pride will not allow otherwise," Danea said. "Kol, why do you wish to avoid what is inevitable? This female will seek confrontation no matter what we do."

"I sense something about her. Something that is not contained within the database." He made a gesture of frustration. "I do not know what it is."

"That's because there's nothing there." I remembered the way Fayne had looked at Kol—like she'd wanted to take a bite of him, too. Another reason to pound her into the quad floor. "Let it go."

"I cannot. We must go carefully from here."

I rolled my eyes. "You worry too much."

"While you are far too reckless," he shot back.

"Enough. One would think you two long bonded." Nalek dropped onto his mat. "We may debate this tomorrow. I am sleeping while I can."

"Sounds like a plan," I said as I went to switch off the optics.

"You are not amusing." Kol came up behind me and switched them off himself, then kept his arm up to stop me. "Come with me."

I followed him into the small lavatory adjoining our room. "They'll still hear if you yell at me, or I ram your head into the bathroom wall."

He wasn't going to rise to the bait—I could tell by the annoying, paternal look he gave me as he closed the door panel. "We cannot follow two paths, Sajora."

"Ah, the inevitable journey philosophy. You forget, Kol, I'm not Jorenian." I leaned back against the cleansing unit and folded my arms. "And I don't need another father, thanks."

"It is always about you, is it not?" He tried to pace, but there wasn't enough space. "Never anyone else. What of Galena? This Fayne is said to prey specifically on the weak and uncertain. Does it not concern you that she may likely attack our ClanSister first?"

Guilt made my face hot. "I'll watch Birdie's back."

"For every moment of every session?" His hand chopped the air. "You cannot."

"I didn't ask her to come here, did I? I didn't ask any of you!" So I was yelling. It was better than shoving his face into the disposal. "If you can't handle it, go back to Joren."

His hand clamped on my shoulder. "We are your kin. We stay together; we fight as one."

"First the religious garbage, then the HouseClan shit. God, you are so predictable." I pushed his hand away. "I don't care what my mother said, Kol. You are not my kin. You're a bunch of strangers who tagged along for the ride, and I'm getting really tired of carrying your weight." I would have shoved past him, but he wasn't giving me any space. "Get out of my face."

Instead of being wise and backing off, he took my tunic in his fists and pinned me against the cleansing unit. "You dishonor me. You dishonor *us*."

I looked down at his hands. "I can do worse."

"I would see you try," he said, so close his breath warmed my face and his scent enveloped me. Different this time—more rain than pine. And hotter, like it had come from a cloud on fire.

The threat should have made my claws spring, but the rage simmering inside me felt different from any I'd felt before. More personal than defensive. Every inch of my skin seemed to be heating up. "Let go."

"We are Jorenian," he muttered, still moving in. "Say it."

"No, we're not." As his heavy body pressed into mine, I understood what was happening. If that wasn't surprising enough, I realized I wanted it. My hands slid up his chest, and I pulled his head down to mine. "Not now."

Kissing had never interested me. Rijor had considered it an unsanitary custom, and no one on the homeworld had ever tempted me to find out why other species liked it so much. I got my first clue when Kol's mouth touched mine.

Nothing had ever felt like this.

The initial jolt shot through me like a triple penalty. The shock of tasting him, his hands moving down my back, our hearts pounding in rapid, heavy sync. The feel and smell of him, the way his muscles tightened and flexed, all so close I felt as if we might sink into each other.

"You smell like jaspkerry and safira," he muttered against my mouth.

The door of the cleansing unit gave way, and we staggered

back together into it. The port sensor snapped on, and a steady stream of warm water poured over us.

All that, and still the kiss went on, changing and deepening until my lungs burned for air and his fingers dug into my hips. The two soggy layers of material separating us became intolerable—I needed his hands on my bare skin, wanted to rip his tunic apart so I could get at him.

Something banged on the other side of the door panel.

"Kol?" Danea sounded tired and grumpy. "What are you doing in there?"

I wrenched my mouth away, dragging in air, seeing the shock echo in his eyes. Water ran down his face, beaded on his dark lashes. I smiled slowly and leaned in to lick a drop from his chin. "You want to kill her, or should I?"

"A moment, Danea." He guided me out of the unit, then stepped away. His eyes and expression went blank. "Your pardon, Sajora, I should not have"—he paused to take a deep breath, then let it out—"I mean to say, *we* should not have done this. It is unseemly."

Unseemly?

I blinked some water from my eyes. "What?" He couldn't be brushing me off. Not after all the heat we'd just generated. Maybe there were some ritual words that had to be said. God knew Jorenians had them for everything else. "What did I do wrong?"

"You did nothing," he said, making another, very formal gesture of apology. "It is not your fault; you do not understand."

"Then explain it to me so I do."

"We do not share a bond." He looked a little uncomfortable. "Sexual intimacy outside bond is forbidden to us."

Mom had told me her people stayed virgins until they got married, but I'd thought she was joking. "You're kidding. Why?"

"Choice is more than marriage. It is the bearing of children."

"Uh, no. Thanks." I grimaced. "I've already decided I'm not the maternal type. Forget about me having kids."

He wiped a hand over the sparkling black stubble on his skull. "We Choose our bondmates for life, and bond only with them."

He was dead serious. "And you can't break this rule? *Ever?*"

"No."

Danea hit the door panel again. "Come out of there!"

I ignored her. "Kol, I hate to be the bearer of lousy tidings, but no Jorenian will ever Choose either of us. We're crossbreeds sired outside a bond. They consider us bloodline pollution. You *know* that."

"That may be so. Still, I hold to the customs of my people." He looked over my head. "I cannot take you."

"Really." I grabbed his soaked tunic and jerked until he was down on my eye level. "What if I decide to take you?"

"Think for moment, Sajora. It is not only a question of Choice." He trailed his fingers over my damp cheek. "We may share the same sire."

My offcoach used to break up fights by dousing two players with a bucket of ice chips. Kol's warning worked just as effectively, and I shoved him away. "Right. I'll remember that. *Clan-Brother.*"

"My heart—"

"Save it." I punched the door release and knocked Danea aside going out. I barely felt the sizzle of pain from the brief contact with her corporeal field.

"What were you doing in there?" she demanded.

"Scrubbing each other's backs." Realizing everyone was still awake and watching me, I went to my mat.

Galena, who had parked her mat next to mine, stared as I stretched out. "Jory. You are all wet."

"I know." I closed my eyes. "I'll get over it."

Despite Kol's orders, I had no intention of dodging Fayne. Hiding from a bully was useless. But she didn't appear on the second level again, and after a few days it became apparent that the confrontation had either scared off all the lesser bullies, or they were content to wait for Blondie to take care of us.

Our training progressed, and by the time we learned how to

function on four hours of sleep, Dursano appeared to award us with yellow bands, which we discovered entitled us to six-hour rest periods. We also earned extra time for meal intervals, which we used to speculate on when we would be advanced while we watched a few bouts in the student quad.

"The targeting trainer says we do not qualify for advancement until we complete seven rotations of training without a single error." Galena sounded glum. "That shall be some time for me, I fear."

"We have yet to enter blade training," Osrea said. "You will improve with time."

"Get out of the way." Cirilo and one of Fayne's bug pals plowed through the center of our group, knocking into Birdie so hard she was thrown to the floor. "Idiot yellows."

I glanced to see Os snatch her completely up off the red-lined floor, preventing her from getting a shock, then saw Kol's arm whip out to grab Nalek's. They used their arms to clothesline Cirilo, who went down, stunned. Before the bug could react, Danea slapped a hand on its back to give it a jolt.

I had my blade out and in Pinhead's face before he could rise. "Which eye would you like to lose? The right, the left, or the middle one?"

"You're dead," he said, grinning as his gray-haired primate friends joined us. "Fayne will see to it."

"Fayne has problems seeing over her footgear." I stepped back as the apes got Pinhead up from the floor. "Don't mess with us, Cirilo."

"Already dead." He started laughing as he and his pals walked off. "Already dead."

After three more weeks of learning to dodge strikes, wrestle, conceal ourselves, and crawl through various simulated environments undetected, we went from yellow to orange bands, and were ordered as a group to report from movement directly to bladework. From that day, the trainer drone informed us, we would be actively using everything we'd learned since beginning second-level training.

"Finally." I touched the hilt of my tån. "This blade probably has rust growing on it."

The class trainer replaced the yellow band around my upper arm with the orange. "Your weapon does not rust. See to it that your skills learned here do not, as well." It scanned the rest of our group. "You are welcome to return to my session for additional practice after bladework at any time. Good luck."

We weren't the only students sent to the class, which was five times larger than any of the others. Some twenty of our peers from other sessions also walked in with us. Already waiting were more than thirty other students, paired off on one side of the room. They were all sparring with various transmutations of their tåns, although none of them were using their blades in the split osu form.

One shrouded figure pacing the perimeter around the fifteen pairs called a halt, and turned to us. "Assume positions on those marks"—it indicated a series of short lines carved into long stretches of the floor—"and kneel."

We stayed together, surrounded by the other rookies, while the experienced students took positions at the very back of the room. As we knelt, the trainer went up to the front and stepped up on a raised platform. He removed his obek-la, revealing a thin, badly scarred humanoid face with large, beautiful brown eyes. Patches of tawny fur grew in an irregular wreath around his brow and chin, which, guessing from the scar tissue, had been burned or scoured off sometime in the distant past.

With the dimsilk shroud removed, I could see he was very small—barely five feet tall. Yet he wore a black dancer's band around his forehead.

Guess size isn't everything.

"I am Bek, a Chakaran male." As he said this, he displayed abbreviated fangs instead of teeth. "You will address me as Bek or Trainer." He studied the four lines of kneeling forms. "There is only one rule here: You will follow my instructions. If I tell you to put down your weapon, you will do so. If I tell you to stand on your cranial case, you will do so. If I tell you to kill, you will do so. If you understand and accept this, stand. If you do not, leave."

After some hesitant glances all around the room, everyone stood.

"Excellent." He pointed to the first line of rookies. "You ten, move to the south corner. Second line, to the north. Third and fourth, you will observe from your present positions. Sparring students, return to your form practice."

The seven of us were all in the third line, so we stayed put.

Bek went to the north and south groups, and ordered them to remove their tåns. "Look upon what you hold. On most worlds, the bladed weapon was created for a single purpose: to open living flesh. Here on Reytalon, the tån was created to train those who wish to kill. It is not the weapon—you are."

The way the trainer said that made a shiver crawl up my spine.

"A novice knows nothing about plying the blade, or the capacity one has for self-preservation. When a blow is struck"—the trainer whipped out his blade and slashed at the nearest student, who threw up an arm, trying to fend off the blow—"the novice parries instinctively, defending himself." Bek turned to the rest of us. "Before you can wield the blade, you must overcome this instinct."

One of the students behind us made a chuffing sound. "We must allow ourselves to be stabbed?"

"No." Bek showed more fang. "You will learn *shahada*—the movements and patterns that allow you to strike first."

The first day in bladework was exhausting. Bek drove the class relentlessly, beginning with our blades in raen-tån form. While we knelt and watched, he had a pair of experienced students demonstrate the two-handed grips required to wield the long sword, then the most basic of cuts, thrusts, and parries. Then we paced in a widely spaced circle, practicing the moves by slicing at the air.

I'd never kept my blade in raen-tån form before, and discovered the holographite increased in density along with size. After a few minutes the weight began to drag at my arms, and within the first hour my muscles cramped. I didn't complain. Some of the other students did, and Bek immediately dismissed them from the class.

"When your trainers certify that your upper-body strength has improved by twenty percent," he told them, "then you may return."

"So there are two rules," I said in a low voice to Galena as I kept moving. "Do what he says, and no complaining."

Bek's ears flicked. "You may complain if you like, Saj. *After* the session is finished."

It was also good to know the Chakaran had ears like a bat. "Will do, Trainer."

We were all ready to drop when Bek finally called a halt, and sent us for a short meal interval. That was when we discovered that having orange bands meant a new, mandatory change in diet, and we had to sit in a special area reserved for students in blade training.

"I think it is protein of some kind." Birdie sniffed at the bland-looking stew we'd been served, and made a face. She was a devoted vegetarian. "Does anyone want my portion?"

"Eat it if you can, ClanSister." Although he wasn't fond of meat, either, Kol dug into his. "This may be the only fare we are permitted for some time."

"Why's that?" I asked. I liked meat, when I could identify it. I didn't know anything, real or synthetic, that cooked up this particular shade of gray.

"Given the nature of this new training, maintaining our muscle mass is imperative. Doubtless this has been designed to aid, and perhaps improve, our physical conditions."

I tasted the stew, which was about as appetizing as it looked. "Or we pissed off someone in the kitchen."

A drone unit brought by a trolley loaded with fresh fruit from a variety of worlds. "When you finish your meal, you may help yourselves." It waited like a stern parent, watching the levels in our bowls.

Galena gave the fruit a wistful glance and picked up her spoon. "This had better help me grow back my feathers."

After the meal, we returned to the bladework room. The experienced students stood around the challenge quad, watching a bout between two blues.

"I wonder what Scar-Face is going to do to us now," I said as we walked through the door panel. Then we stopped, and looked up with everyone else.

Dozens of bodies had been hung on cords from the ceiling, and swayed gently over our heads. They were swaddled in plain white fabric from head to foot, and the cords had been tied tightly around their necks.

Osrea muttered a short prayer, then added, "I did not think the stew was *that* bad."

"They're not real." I reached up and felt the foot of one of the "bodies." "Stuffed."

"These are practice targets," Bek said. He went to a wall console and pressed something, and the cords began lowering the figures. I saw how the cords were attached to alloy poles fitted into scrolling tracks on the ceiling. He flicked another switch, and red lights glowed beneath the white fabric shrouds. "The lighted areas are kill zones. Use the forms you have practiced today to strike these areas. Take your line positions."

We resumed our places, with a practice form dangling in front of each of us. I thought mine looked a little like Fayne.

"Remove your blades and transmute to raen-tån." Bek looked around the room, and when everyone was armed, he added, "Begin."

Almost at once, someone yelled. I found out why when I looked away from the cut I was making and my tån went outside the red zone. A healthy jolt traveled up from my blade into my arm, making me yelp.

"If you do not strike the kill zones, it will cost you," the Chakaran announced.

I rubbed my forearm. "Now he tells us."

We were all more careful with our blows from there, but still, every couple of minutes someone hissed or squealed in pain. Since the targeting trainer had emphasized the importance of going for a cardiac strike, I concentrated on cutting a nice big hole through the red chest zone.

"Do not hesitate!" The Chakaran stepped between a novice and the stuffed form suspended from a pole in the ceiling, and took the tån from the student with ridiculous ease. He struck a single blow to the novice's chest. "Now I have your weapon, and you are dead. Students, halt your practice and attend me here."

We stopped stabbing and slashing at the forms hanging from our poles, and gathered around Bek.

"During your targeting training, you were taught the three objectives when confronting an armed, vigilant opponent. Saj"—he nodded at me—"recite them, in order."

Luckily, I still remembered them. "Disarm."

Bek swept the student's tån to one side, cutting off the form's hand, then reversed the move and cut off the other.

"Disable."

With two more strokes, he amputated the form's stuffed legs.

"Dispatch."

Bek thrust the tån into the chest target indicating the cardiac organ, then stroked the blade across the form's neck. What was left of the body collapsed onto the floor, leaving only the head hanging from the pole.

There was a moment of silence as we all stared, first at the decapitated form, then at the little Chakaran.

"When you attack, there can be no pause, no consideration, no discontinuity. A split second of indecision creates opportunity for retaliation." The trainer handed the blade back to the now-pale student and called for a replacement form. "Resume your practice."

As we went back to hacking up the forms, the ceiling rods began to move, making targeting even more difficult. Out of the corner of my eye, I noticed Bek weaving his way through the lines, watching each of us for a short period before moving on to the next student.

"You have used a blade before," he said when he got to me.

"A knife, not a sword." I went for the disarm cut and managed to almost amputate my target's left hand. "Will we really have reason to use the tåns in raen form, Trainer? Sword fighting went out on my world about a thousand years ago."

"The blade is silent; powered weapons are not. Many civilizations have forgotten how to defend themselves against long blades, and the raen form would best serve you against them. It is especially effective when attacking multiple-armed opponents, for example." He made a circle around me. "You are part Terran, are you not? You and the other crossbreed male?"

"Yep." I grunted as the tip of my blade ventured outside a red zone, and I got zapped again.

"To date, I have taught only three Terrans to dance." Bek evidently found that amusing, from the growled laugh he uttered.

"Well, between us we make a whole Terran," I said, glancing at Kol. "Count us as four."

"I will," he said as he moved on. "If you both survive."

"Look." Renor nodded toward one of the two fighters climbing into the quad. "A challenge is about to begin."

Their blurry figures made my eyes narrow. "Hey, how come they get to wear dimsilk, and we don't?" I demanded.

A passing red student paused for a moment to tell me, "All students don dimsilk for challenges."

"I know that one hanging the red band in his corner," Ren added when the student moved on. "He is a Threkr, and will prevail."

"Why will he win?" I studied both fighters, who were wearing dimsilk. "They look pretty evenly matched."

"The Threkr are a powerful equine species, and none of his sparring partners have prevailed over him yet." Ren's cheek flashed. "Including me."

A hoverdrone descended to initiate the bout, and I glanced over at Kol. Since our own private wrestling match, he had gone to great lengths to stay away from me. Which was fine. Most days I still felt like stabbing him in the heart.

I cannot take you.

As if I'd ever want to tie a three-hundred-plus-pound Jorenian around my damn neck for all eternity. I turned back to the quad in time to see the drone rise and the two challengers approach each other.

The Threkr's opponent seemed unremarkable, except for the

green band it hung in its corner. "Do you know the other one, Ren?"

"No."

Had someone been taking bets, I'd have put my credits on the horse-guy at first. But a few minutes later I started to wonder just how much of an advantage strength really was.

The two fighters began the bout with their blades in raen-tån form, the length and weight of which required a two-handed grip on the hilt. I should have been analyzing their movements as the pair thrust and parried and slashed, but the lethal elegance of the dance mesmerized me.

"A well-matched pair," Danea murmured beside me. "The Threkr will not prevail so easily, I think, Renor."

The green-band fighter transmuted its blade through the thion- and rangi-tån forms, always advancing, effectively taking the offense away from the Threkr. Gradually the circles they danced around each other grew tighter, and their tåns shorter, slimming down through jyan-, shou-, and elok-tån forms.

"They're going to osu-tån," a student near me said, his voice thin with excitement.

"They don't allow two-blade fighting on the second level," someone else told him.

The first one made a sound of contempt. "Only in blade-work practice. There are no rules in the challenge quad, except to survive."

Kol moved forward, drawn to the side of the quad as though fascinated. His eyes followed every engagement of the blades, and I saw his fingers curl around the hilt of his own tån.

He wants to jump in. I could feel the intensity of his concentration, and wondered why he was so caught up in this particular fight. We'd seen others he'd walked right past.

I looked back at the challengers. Green was moving faster than red—much faster—and seemed to be everywhere, slashing mainly at its opponent's chest. The hoverdrone, which maintained a safe distance above the clash, made a funny sound, and everyone around us took out their blades and struck them together.

"One hit to the red challenger," the drone announced.

"Why does everyone have their blades out?" Osrea asked me.

"Maybe they don't want that green coming after them."

"We count the hits by clapping tåns," one of the other students told me. "Green only has to strike one more before red will have to capitulate."

Visibly tired, the red challenger tried to make up with a burst of fancy thrust patterns. Green nimbly danced out of the way, darting under red's guard to slash at his dimsilk. Part of the fabric peeled away, and I saw a series of crisscrossing wounds streaming with milky-looking fluid.

Then the green challenger transmuted its tån to split, and every student around us seemed to freeze. The red fighter staggered back, confused, and green chose that moment to strike again.

The drone dropped down to hover just between the two fighters. "Two hits to the red challenger. Red is advised to withdraw from the challenge now."

The sagging challenger pulled his obek-ten from his head, revealing a sweaty, pale face. He moved toward the green fighter, still holding his blades in each hand, but it was obvious he didn't mean to attack.

The green fighter leaped into the air, flipping over and landing behind the red challenger. With brutal efficiency, it drove one blade into red's chest implant.

The Threkr stiffened, his blades dropping to the quad surface; then he fell forward, landing with a heavy thud. He didn't move again.

"That is not fair!" a high-pitched voice yelled, and Galena ran to join Kol at the side of the quad. "He was through fighting!"

"Three hits to the red challenger." The drone rose and buzzed away. Two trainer drones entered the quad and dragged the student's still body away, leaving a splattered trail of milky blood behind them.

The green challenger walked over and crouched down at the side of the quad. "You wish to challenge me?"

"You murdered him," Birdie said, though not quite as loudly. "He meant to capitulate; we all saw it. How could you attack him like that?"

"He approached me, and I defeated him." The fighter pulled off its obek-ten, and shook out her pale hair before she smiled at Kol.

Fayne.

"What about you?" she asked Kol. "You look ready to die today. Shall we dance, breed?"

Before I could move an inch, Nalek and Osrea grabbed me by the arms. "Let go, you guys. I just want to have a talk with the little shrimp."

"No, Jory," Nal warned me. "Let Kol deal with her."

Os leaned close. "If she kills you, you'll never balance the tally."

"Even the score," I corrected him through clenched teeth.

"Whatever."

By that time Kol had pushed Galena behind him and was speaking to Fayne in a low voice. Whatever he said, it made Blondie's expression turn smug. He nodded to her, turned, and guided Galena back to us.

"We will report for our next training session now." Kol pointed to the stealth chamber.

"What did you say to her?" I demanded, jerking my arms free of Nal's and Os's grips.

"I offered her a compromise. She accepted."

Sparky frowned. "What sort of compromise?"

"She will not challenge any of us. In return, when I advance to the third level, I will fight her. Should I prevail, she must leave the Tåna." He glanced at me. "Should she prevail, we seven will leave."

At that moment I could have killed him. Easily. With no regret. "So if she wins, we'll be able to personally escort you back to Joren. In a body bag. How nice."

I stalked off to class.

CHAPTER THIRTEEN

"THE TRAVELER JOURNEYS WHEN IT IS TIME TO FOLLOW THE PATH, NOT WHEN IT IS CONVENIENT TO DO SO."

—TAREK VARENA, CLANJOREN

Blade training made everything that had happened before seem like kid's stuff. When Bek wasn't pushing us through exercises on form, he constantly nagged us about ignoring our instincts for self-preservation, until I thought I'd stab him in the heart myself just to shut him up.

From there, days blurred into weeks. We practiced for hours, stopping only to eat our tasteless meals, then went back and practiced until we were ready to drop. We barely spoke to each other when we were dismissed from bladework, too tired to make even an attempt at conversation. Most nights we filed into our quarters, secured the door panel, and dropped on the nearest unoccupied mat.

I was usually unconscious before my face smacked the floor.

At last the Chakaran felt satisfied that we'd adequately gotten over our defensive instincts, and that was when he put us to work on blade form.

"You will master first the raen-tån, the long sword." Bek held his up, and again I noticed the difference between the trainers' blades and ours—the black alloy glittered with keen, lethal promise. "This is the longest and heaviest of blades. It presents certain advantages and challenges during the *shahada*. We will explore these aspects as they apply to your potential success or failure when engaging a target."

Through each subsequent session, Bek stepped up the pressure, demanding more every day. From the very basic attack forms he began adding nuance after nuance, drilling us on each until we were able to execute them without error. Our inventory of specific moves went from three to thirty to more than three hundred. We perfected our techniques through dogged repetition, but the trainer insisted there was more to the dance than proficiency.

"The blade must belong to you, as much as another hand or limb," Bek would say, jumping immediately on anyone showing the slightest awkwardness with the tån. "It must be that when you move, it moves. No thought, no self. You become a single presence."

He didn't like heavy-handed hacking, either. "You do not wield an implement. In the hand of a true dancer, the blade acquires a soul."

The picky Chakaran also hated errors with a passion. "Anticipate the strike. See it penetrate the kill zone before the blade makes the initial incision. Change the direction of your blow the moment you see it will fail—turn the downward strike upward, the right cut to the left."

Gradually we adapted to the rigors of blade training, from bearing the heavy weight of the raen-tån for hours to the enforced, dismal diet of synthetic protein. I noticed that we were all dropping weight fast, even Galena, who didn't need to.

Still, our bodies and physical conditions improved despite the rigors. Shedding what extra body fat we possessed allowed sleek, hard muscle and sinew to emerge. Osrea's exocartilage plates shrank along with his bulk, but they doubled in thickness. Though painfully thin, Galena seemed to be getting a little taller, and her winglets, which I kept nagging her to stretch instead of holding them flat against her back, were growing and expanding. And Nalek's physiology seemed to thrive on the requisites, until his ridges literally rippled with power whenever he moved.

While bladework helped us to build our bodies and develop better physical tolerance, it exposed some significant problems, too. I started noticing them as I watched Bek evaluating us during the sessions. The same mistakes kept cropping up over and over.

Galena, by virtue of her size and body weight opposed to the expanding wings on her back, had trouble maintaining balance. She regularly overcompensated and buckled under the weight of the blade. This made her miss kill zones, sometimes so often Bek threatened to send her back to the basic classes.

Danea could not spar with anyone without jolting them, so the trainer ordered her to wear insulating thermals under her garments. Losing the advantage of her corporeal field made Sparky more cautious and prone to lock up her blade during sparring, which the Chakaran read as hesitancy.

"Advance; do not counterparry!" He pushed her back from her opponent. "Roll the blade under and break the contact!"

Nalek had the same problem, for different reasons. When we surpassed the forms and began sparring with each other, he ended up inevitably on the defensive.

"A dancer does not lack aggression," the Chakaran told him. "Find yours, Nal, or you will end as an overlarge skewer of dark meat."

Osrea suffered from extreme clumsiness. Half of his upper limbs remained bound during training (because we trained with one or two blades at a time, species with more than two upper-body limbs were required to wear special tunics that kept all but one pair of limbs pinned underneath the fabric.) Gradually I realized Os's situation would have been like me trying to fight with one arm tied behind my back. He had dropped his blade more often than anyone in the class.

Renor, like Danea, proved too dangerous to be allowed to practice without protection for his sparring partners. His crystalline hide inflicted more wounds than his tån. Bek came up with a thick, padded garment for him to wear under his outer garments, but it slowed his reactions, and he also missed too many targets.

Bek often examined his target form with disbelief. "Ren, are your eyes not functioning?"

I was able to keep up with the demands of the class, but my knee was beginning to give me trouble, especially on the more complicated pivoting moves. Thgill's prediction about a serious percussion injury kept me skittish. As a result, I avoided or faked my way through certain techniques, and occasionally incurred Bek's wrath when he caught me.

"You cannot skip your way through the *shahada,*" he told me after stopping a practice match I had nearly won. "Every step is essential for your success and personal self-preservation."

"They're just shortcuts," I said, feeling a little self-righteous. I hadn't made any mistakes, other than gliding through Bek's more difficult whirling moves. "I'd have won; isn't that the point?"

"Winning is always the point." He took out his blade. "Now spar with me." Ten seconds later he had me on the floor, pinned under his footgear, his blade at my throat. "Any protector with adequate training would spot these openings you create in your guard with your *shortcuts.* And use them to kill you."

Only Kol seemed to have no problem at all. He handled the tån as though born with it attached to his arm, and sharpened his attack skills to the point of never losing a sparring match. The Chakaran used him more as an aide for a short time, then sent him to spar with the more experienced students while we repeated form practice. It was during one of those sparring matches that he accepted his first challenge.

Several of Fayne's cronies had been recycled back to second-level training, and Kol had the bad luck to be matched against one of the bigger, meaner ones late one session. He defeated him with embarrassing ease, which infuriated the larger, heavily clawed male.

"Orange does not beat red! I challenge you to the quad!" the loser shouted. "Then we shall see who prevails!"

As usual, Bek didn't stop the training, but ordered Kol and his challenger out to the quad. I couldn't concentrate, and took two jolts for missing kill zones before the Chakaran gestured for me to step onto an insulator pad away from the rest of the class.

"Saj, you tempt me to send you back to white." He glanced at the closed door panel between us and the quad. "I will give you some advice: Dancers do not allow themselves to be distracted. By anything."

"I know." I studied my footgear for a moment. "Bek, do you have a family? A mate, children?"

"No."

I met his flat gaze. "Did you ever want some?"

"No. They would be made to suffer because of me." Absently he touched a burn scar on his brow. "What has this to do with your inadequate performance?"

"I never thought I'd have a family, but this clan has become mine." The hilt of my tån dug into my palm as I put my entire training on the line. "Send me back if you want, but I need to be out there with him."

"I see." Bek's head turned toward the third line, where my other five headaches had stopped seriously practicing and were simply going through the motions while watching us. "The others feel the same."

Thinking he might send everyone back, I panicked. "No, it's just me—"

He held up a paw. "I have observed you seven too long to be deceived by your desire to protect the others." He gestured for the rest of the gang to join us. When they did, he said, "I have never encouraged relationships among my trainees. It creates too much conflict with the work to be done. However, I can see how strong this group's interdependence is, and I doubt I can do anything to eliminate it." He nodded toward the door panel. "You are dismissed from training to observe the bout. Return when it has finished."

"Bek does not strike me as particularly empathic," Danea said as we hurried out to the quad. "What did you tell him?"

"That you guys were more important than my training." I gave her an irritable look. "I lied."

She hmphed. "I thought as much."

Someone had spread the word to the third level, because there were dozens of red, green, blue, and purple bands crowding the quad. As we pushed our way through to the front, I saw that Kol and Fayne's buddy were already well into the match, their tåns flashing in the shorter jyan form. Nalek used his bulk to shoulder a space for us against the center of one side, but Kol was too busy countering moves to notice our presence.

Both of them were in dimsilk, and moving so fast it was hard to see the individual moves, until their blades locked and created a brief pause. Kol's opponent broke the lock, swiping at Kol's abdomen with a choppy lateral cut and slamming a fist into his head from the opposite direction at the same time.

Kol's obek-ten went flying, but instead of being thrown off balance, he used the opportunity to get under the red's guard, attacking with a savage thrust to his chest. Not anticipating the hit, the red staggered back against the ropes.

The hoverdrone descended. "One hit to the red challenger."

Everyone clapped their tåns as Kol scooped up his helmet. He saw us and made a quick gesture I'd never seen before.

"What does that mean?" I asked Galena.

"Guard the House," she said. "It means we are in danger and must watch out for each other."

The press of students in back of us suddenly took on a more ominous quality, and I moved behind Galena. "Sparky, clear a space, will you?"

Danea also took a position at the back of our group, energy spiking her short hair as she scanned the surrounding faces. Everyone within five feet abruptly shuffled out of striking distance.

In the meantime, Kol was back in the fray, transmuting his blade down to match his challenger's shou-tån, eliminating half the previous fighting distance between them. The fight was just as fierce and fast as before, but I noticed the challenger had quit using the more sophisticated combinations and was battering Kol with sheer brute power. The force of his thrusts and cuts locked up their blades more than once, but Kol's guard never faltered.

"One of them will snap a blade soon," someone crowed.

"Is that possible?" I asked Renor.

"Given the nature of the tån, yes." His crystalline face remained impassive. "A broken blade is counted as a hit."

The red danced back, out of reach, and turned away, presenting his unguarded back as he looked up at the hoverdrone. For a moment I thought he might call it quits. Then he lunged at Kol, his tån separating into osu form, slashing down at Kol's chest from right and left angles.

With his blade still in shou form, Kol couldn't match the move, and everyone knew it.

Then something impossible happened.

The challenger's two blades screamed as Kol's met them—he had somehow anticipated the attack and shifted to osu-tån. He met the lunge with such force that a bone cracked, and the red shouted in pain. Kol followed through from the other direction with a quick, efficient thrust, and buried his blade in the challenger's chest for the second time.

I'd never seen anyone react that fast.

"Two hits to the red challenger. Red is advised to withdraw from the challenge now."

The challenger did just that by dropping his blade and collapsing onto the quad floor. His broken arm hung useless as he stared in complete disbelief at Kol. "You did not see. You could not know."

Kol sheathed his blades. "There was no other reason for you to abandon your guard as you did." He held out a hand to help the challenger to his feet, and after a moment of hesitation, the red took it. "It was a good bout; I learned much."

A good bout. He'd learned a lot. He'd nearly gotten himself butchered, and he liked it? I made a mental note to personally club him over the head at the next available opportunity.

I wasn't the only one who thought the man was deranged. As the medics climbed into the quad, the challenger cradled his arm and faced Kol.

"You will regret this more than I, crossbreed." The red made a harsh sound. "I have but to recover and retrain. The entire Tåna will seek to challenge you now."

"Let them," was all my lunatic ClanBrother said.

Bek appeared as Kol left the quad. He pointed to a dark figure waiting by the targeting room. "Uel wishes to speak to the seven of you," the Chakaran said. "Attend him now."

* * *

We filed into targeting, and took up positions on our marks. The Blade Master dismissed the trainer drone and went to the room console.

He didn't mince words, either.

"Information affecting your homeworld has been received." He pulled up a star chart and projected it as a dimensional image into the center of the room. "Combat fleets from both the Allied League of Worlds and the Hsktskt Faction have pushed forward through the Goldokis Quadrant and now threaten to engulf Varallan space within the cycle."

Which put Joren directly in the path of war. We looked at the projected swarms of ships approaching from either side of the quadrant. There were thousands of them.

"Why tell us?" I asked.

"It is the consensus opinion of my staff that none of you will survive training." The Blade Master shut down the projection and turned to us. "I have decided to give you the option of leaving Reytalon so that you may return and defend your homeworld."

"You want us to quit?" I couldn't believe it. "Just like that?"

The obek-la covering his face blurred everything but his chilly voice. "I offer you an alternative to death."

"Why?" I stepped off my line, something that would have gotten me a good jolt from a trainer. "You aren't making this offer to anyone else, are you?" He shook his head. "Why just us?"

"Jory." Nalek looked miserable. "Our homeworld is in danger. Our place is there."

"Your homeworld, big guy. Not mine."

Kol also moved off his line, walked up to Uel, and moved around him in a slow circle. "I would like to hear the answer to Sajora's question."

"Not all of us are as skilled as you, Kol Varena." Osrea folded his arms, making his exocartilage plates grate. "What he says is true. We will not survive in the quad."

Renor looked at the door panel. "The albino Skogaq directs her comrades to challenge us."

"We cannot help our people if we are dead," Galena whispered.

"Why are we debating this? Let the Terran stay if she wishes to embrace the stars." Danea's hair practically stood on end. "*Our* homeworld is in danger. *We* go back."

"Sajora will have her answer now"—Kol continued circling Uel, his hands on his blade hilts—"or you will dance with me, Blade Master."

Uel held out his gloved hands. "Calm yourselves." He turned to me. "Jorenians have a reputation for seeking revenge on those who harm their kin. Revenge is why you are here, is it not?"

"We could be here because we're bored," I told him, and wondered if the dancer from the *Chraeser* or Uzlac was responsible for what Uel knew about us. "What does that have to do with it?"

"Official inquiries have been made regarding your whereabouts. One from a council member's bondmate. We use certain methods to preserve our anonymity, but it is only a matter of time before your people track you to Reytalon." He turned back to Kol. "Our training techniques already violate your HouseClan laws about harm inflicted on kin."

Nalek shook his head. "We can shield you before our Houses, and no retribution will be taken."

Uel nodded. "And what will happen if your people arrive and find you have perished here?"

"They will declare ClanKill on you and the trainers," Kol said.

A little silence fell over our group as we all imagined what the Tåna inductors and trainers, all of whom were blade dancers, would do to a group of enraged Jorenian warriors.

"We're not going to die here." I looked steadily at Kol. "Joren isn't going to war tomorrow. We're going to stay and finish our training."

"Joren is in danger!" Danea shouted.

"Not yet." For once I didn't feel like snarling back. "And if you go back when it is, what will you do? Shock all the Hsktskt to death? I don't think you have enough juice to do one, Sparky."

"We will join our HouseClan warriors; we will. . . . They will have to allow us . . ." She trailed off as Kol shook his head. "Surely during war, Kol. Surely they would not detain us again."

I frowned. "What do you mean, again?"

"It happened three revolutions ago, during the last attack on Joren," Osrea said. "I was sent to a security facility and kept there for the duration."

"My ClanLeader feared I would betray our HouseClans in battle," Nalek admitted. "Inadvertently, of course."

Birdie looked like someone had given her a full-swing face slap. "My ClanMother said it was for my protection. That I might be captured . . . harmed . . ."

"Did they do this to you, Kol?" He nodded. I already knew Ren had been stuck in a cell. No wonder they'd wanted to come with me. If they'd stayed on Joren, they'd have ended up in prison. "Oh, yeah, they're definitely going to welcome you back with open arms this time. Hand you weapons, send you out to battle for the HouseClans, et cetera." I made a rude sound.

"You say you will complete your training, Sajora," Uel said to me. "How will you keep yourself and the others alive?"

I thought for a minute. "We don't accept any more challenges. Doesn't matter who makes them; we turn them down. That'll keep us out of the quad. After sessions, we'll train by ourselves. Focus on dealing with our handicaps. Watch one another's backs." I turned to Kol as he came to stand beside me. "We can do it, if we work together. That's how we got this far. You know we can, Kol."

"It must be a commitment we all make." He eyed Danea. "What says the HouseClan? Do we stay or go?"

"I am not eager to return to an isolation chamber," Renor said. "I will stay with you."

"I did not enjoy being detained before." Nalek looked at his footgear. "I stand with our House."

"They have likely filled in my dugout," Osrea said. "I do not feel like digging another. I stay."

Galena stroked the new feathers covering the top of one wing. "I miss my ClanMother, but . . . I will remain with you."

Danea stepped between me and Kol, still seething. "You persist in annoying me, Terran."

I smiled. "I haven't even warmed up yet, Sparky."

Her purple lips thinned. "Kol, know this: If anyone kills Sajora, it will likely be me."

"I trust in your restraint, Danea."

She nodded. "Then I stay as well."

"That makes it unanimous." I glanced at Uel. "What say you, Blade Master?"

"There are many ways to progress through training in the Tåna. In some cases, courage and determination count as much as acquired skill." He tossed a bundle of red bands at Kol, who caught them reflexively. "You seven are advanced to red and will move to third level, effective immediately."

"It could have been another trick, like the prison and the magma pit," I was saying to Ren as we left our new quarters the next morning. "Uel could have come up with that story about the war approaching Joren, just to see how we'd react."

"It is unlikely," he said. "Such a test would be given to all students, not merely we seven. Third-level trainees are given access to the Tåna's database, so the facts can be easily verified."

We had been directed to take one of the restricted corridors to a lift, which rose silently into the upper level. When the door panels parted, we found out why the students called the area used for the final phase of training "the bubble."

Third level was enclosed by an enormous, transparent dome. Within the curving, transparent walls, individual and group training rooms had been built and stacked to provide each with a view of the sprawling arena below, and the icy, lifeless surface of Reytalon beyond the dome. Yet it wasn't sight of the outside world that silenced us.

A small army could have comfortably sparred within the enormous center quad. And a large army of color-banded trainees already milled around the base of its platform.

"So many," Galena said on a released breath.

Kol studied the four-cornered platform. "They must practice in groups, or confront multiple opponents."

Above us, clusters of trainees began moving in and out of the session rooms, some descending from higher tiers on open lifts, others pacing the red-lined floors to watch what was happening below them. We seemed to be drawing the most attention, from the number of stares directed at us.

"Uel has made arrangements for us to have our own space, after the regular sessions." Kol inspected a trio of green bands murmuring as we passed. "No one is to be alone. From this point on, we stay together, or in pairs."

"Did you notice?" Nalek swept a hand from side to side. "There are no reds here but us."

"You are the first red bands to be advanced to the third level." Dursano appeared in front of us, and gestured to a large room on the west side of the quad. "Follow me."

We weren't given a tour this time, but a briefing. The third and final stage of training included similar classes to the ones we'd taken before, but this time we would battle other students rather than forms, drones, and laser targets.

"We go totally live, then." I looked through the viewer at the quad. "No more safeties."

Uel turned his shrouded face toward me. "You may still leave Reytalon, if you wish. We will provide transport to Joren for the seven of you."

"No." That came from Galena, surprisingly enough. "We stay."

We were dismissed to begin our new targeting class, but I lagged behind for a moment. "Blade Master, may I have a private word with you?" He inclined his head, and I glanced at Renor, who was playing my shadow. "I'll meet you outside."

When we were alone, I gestured to his obek-la. "Do you ever take that off?"

"No."

"Okay." So it was talk to the mask. "I met a blade dancer on my way to Joren. He owns the Rilken gun runner *Chraeser.*"

The Blade Master walked over to the console and cleared the viddisplay. "I know of him."

"He was the one who told me about Reytalon." I paced, trying to choose words that would get me what I needed without offending Uel. "I need to find a human named Kieran. He was born on Terra but he spent most of his life out here, in space. The dancer said you knew him."

Uel kept his back to me and said nothing.

"I really have to find him." I didn't like asking for help, but I didn't like the idea of chasing a raider around the galaxy for the rest of my life, either. "When I graduate, will you help me locate him?"

The Blade Master turned around, and although I couldn't see his face I sensed he was completely focused on me. "Why do you wish to find this raider, Sajora?"

"He'll be my first professional kill." Uel didn't respond to that. "Do you know him? Will you help me?"

"I admire your goal, but I cannot assist you." He went to the door panel.

"Just tell me where he is."

He left without replying.

Renor escorted me to targeting class, where he was temporarily assigned as my sparring partner. The trainer told us we would be moved around until we ended up partnered with whoever best matched our abilities. I noticed Kol had been matched against Danea, and although they fought with lethal concentration, she couldn't keep up with his speed, and he kept pulling his thrusts back at the last moment.

"You are not concentrating," Ren said after I missed him for the third time. "You leave too many openings in your guard, like so." His blade flashed as he thrust it toward my midsection.

What happened then was a revelation.

Without thinking, I shifted my weight to my back leg and slid my lead foot back, curving my body to the side to create an empty space without retreating. At the same time, I used a reverse cut with my left-hand blade and hooked him by the wrist, slipping under his guard. With my right elbow, I caught the inside of his left arm and forced it out, carrying through until my right blade stopped an inch above his implant.

All of that took about two and a half seconds. Pure reaction without thought—what all the second-level trainers had been harping about—was a lot like running for the zone, I realized. It was possible to fight without thinking about it.

I let my gaze drift, and saw the Blade Master watching us. *What's he doing here?*

Ren sighed, but remained locked in position with me. "That was interesting."

"I could have killed you," I said, feeling a little belated guilt.

"I know." Finally his cheek glittered. "Do it again."

The trainer, a wide-bodied alien who left his ruddy-skinned head uncovered, appeared beside us. "A lateral inside response to the female's move may have saved your glassy hide. And you." He gave me the once-over. "You have found the blade in yourself, and yourself in the blade, have you not?"

It described what I felt better than I could have, so I nodded.

"You will be paired with another." He gestured for us to move apart, then pointed to Kol and Danea. "Trade partners—yellow hair, oppose the crystal one."

We switched, and for the first time I became acutely aware of one fact—besides the one time on Joren, Kol and I had never really sparred with each other. Not with blades.

The trainer made it worse by instructing us to assume position on the exhibition platform, at the front of the class. "You may demonstrate the various lateral closes for these others. Outside and inside, in osu-tån."

I walked up to the platform and faced my ClanBrother. "I spoke to Uel about Kieran," I told him in a low voice.

"A target attacks you," the trainer said. "Your response is first outside lateral form, avoid, control, hold. Execute."

At the command, Kol stepped forward and thrust his blade toward my rib cage. "What did he say?" he murmured.

I stepped out to the left, just enough to let his blade pass by me so I could clamp his forearm between mine and my body. "He knows him, but he won't help me."

Kol countered my pin by bringing his left blade up toward my throat. "It is possible they are comrades."

"Maybe." I lifted my chin up and back while I parried with my right blade, forcing his left hand away from my face. "The moment I said Kieran, I had his full attention."

The trainer stepped onto the platform. "Stop. Resume attack position."

We lowered our blades and stepped away from each other.

"The instinctive reaction to the pin and blocked hook will be to step back," the trainer said. "Respond to the retreat with an immediate attack, and you complete the kill."

"But neither of them retreated," someone said. "If the target responds as they do, there is no space, and no opportunity to strike."

"An excellent point—yet one must yield, even if it is only to exhaustion." The trainer turned to us. "You will spar on exhibit until one of you retreats."

As soon as he left the platform, we stepped forward and went at it again. Like me, Kol tried to stay as close as possible, eliminating any distance for maneuvering. We fought for control of each other's blades, gaining it one moment only to lose it the next. We went through all of the lateral outside forms, then the inside techniques, until it became apparent neither of us would yield to the other.

I knew I'd be the one to lose. Kol was much bigger and stronger, and the sheer physical stress of enduring each impact while countering his moves was starting to wear me down. Still, I didn't *want* to lose to him, to be the one the trainer pointed to as the failure. Not in front of the Blade Master, who was still watching us.

No, I thought, recalling other times I'd caught glimpses of him watching, when Kol hadn't been present. *He's watching me.*

That realization seemed to push me through fatigue into the same strange, detached place I discovered sparring with Ren.

Having watched him fight all these weeks let me anticipate most of Kol's moves. He had an aggressive, devious personal style that depended as much on strength behind the blade as the element of surprise in front of it. He seemed to be reading my mind, too, for he answered my moves with uncanny speed.

Time dwindled and disappeared as we danced, always close, our blades moving in perfect synchronization. We were breathless and covered with sweat when the trainer finally called a halt to the match.

"This will continue tomorrow. You are dismissed for meal interval."

I frowned as I sheathed my blades—it felt as if we'd just gotten started—and then I saw the wall panel and blinked.

Either someone had advanced the chronometer, or Kol and I had been sparring for two hours.

CHAPTER FOURTEEN

"No paths cross without purpose."

—Tarek Varena, ClanJoren

Everyone remained noticeably quiet as we sat down to eat our flavorless meal. Kol seemed preoccupied by his thoughts, but the rest of the clan kept sneaking suspicious looks at both of us and each other.

"Something wrong?" I finally asked. "Other than the food, I mean?"

"That exhibition match," Nalek said. "The way you and Kol sparred—I have never seen the like."

"Yeah, so?" I finished my stew and grabbed what resembled an apple from the fresh cart. "It's not like we rehearsed it."

Osrea made a rude sound. "There was nothing natural about it. No one fights that way, not against each other."

"Kol would have prevailed," Danea said.

"Maybe." I took a bite of the near-apple and instantly spit it out. It tasted more like a persimmon crossed with a lemon. "Maybe not."

Renor wiped his glittering face, which turned his napkin into a handful of shreds. "You moved without warning, and yet you each knew each other's tactics."

Galena's iridescent eyes moved from my face to Kol's. "Perhaps they can read each other's thoughts."

"Psychic fighting. Right." I laughed. "Look, we've just watched each other spar long enough to know how we're going to move. We've paid attention; that's what you're supposed to do."

"There is an alternative explanation, but . . ." Nalek made an embarrassed gesture.

"No." Kol abruptly emerged from his reverie and joined the conversation. "It is as Sajora says. We have grown familiar with each other's fighting style."

Everyone went back to eating without another word.

When we left to return to our next session, I fell in beside Kol, who was lagging behind the others. "Mind telling me what that was all about?"

"I do not know what you mean."

Again with the Jorenian attitude. "Don't play stupid alien with me; I know better. Nalek's other explanation—what, exactly, was he talking about?"

Kol made a gesture that joined both hands briefly. "He suggests we share a warrior's bond."

We passed through the door panel for stealth training and found ourselves stepping onto an icy, windswept plateau, remarkably similar to Reytalon's surface conditions. Only this world had a giant sun that veiled everything with blazing blue light.

"Welcome to Akkabarr." A trainer tossed us a couple of white fur parkas and trousers lined with soft, dense wool. "Don these garments and assume positions behind the drift ridges on the north side of the simulation."

"What's our objective?" I asked. "Besides turning into big icicles?"

"Kill the enemy and stay alive." The trainer walked off.

As we put on the cold weather gear, I shivered and squinted at Kol through the whirling snow. "Tell me about this warrior's bond thing."

"A Choice made during battle, formed without conscious intent. We are not at war." He jerked the front of his parka closed. "Even at such times, it happens but rarely."

"I see." No, I didn't, and he obviously didn't want to talk about it. I'd have to corner Nalek and get another battle/love lesson later. I noticed some robed forms materializing to the south of us. "Looks like the trainer's getting ready to initiate the sim."

"Hurry." He gave me a push, and we ran toward the high banks of snow.

Flying crystals, as fine as diamond grit, nearly blinded me as we raced against the cutting wind to take up our positions. Everyone was huddled together on the blue ice behind the drifts, cringing from the snow-laden gusts and glaring light.

"Adjust your footgear straps, and be cautious with your blade grips," Kol said. "Blister fluid freezes in these temperatures."

"What about frostbite?" I said through chattering teeth.

"You only feel pain after it thaws."

As we watched the advancing rows of simulated Akkabarran warriors, our parkas froze on our bodies, making them into stiff, icy boxes that hindered movement. I lost all sense of smell, and could almost feel my heart slow as it struggled to pump blood to my freezing extremities. If we didn't get moving, in a few minutes we'd all be dropping with acute hypothermia.

I glanced up and saw a recessed space in the snow. A window of some kind—and the Blade Master stood on the other side of it, looking down at me.

What the hell does he want?

The other students began retreating to steeper slopes, trying to scale the icy ramparts to evade the Akkabarrans. Several were shot on the way and fell to the ice, stunned and writhing.

"We get stabbed *and* shot?" I muttered. "This is new."

"There is no escape. We must go out on the pack ice," Kol said, reading my mind. "Use ground cover until we can flank them from the east and west."

"What ground cover?" I took a peek at the enemy, who were only a few hundred meters away. They were armed with pulse rifles and curved spears. "They're going to spot us the minute we step out into the open."

"They will not, if we do this." Nalek took off his parka, turning it inside out. Unlike the weather-repellent white exterior fur, the warm blue lining blended with the snow.

Osrea grimaced as he did the same. "We shall freeze before we're hit."

A light went on inside my head. "If we kill all the sims, the trainer has to end the session. It's not just surviving the cold; it's timing the attack. We wait too long, they'll split up and take longer to eliminate."

"I agree; we must be swift." Kol split us into two groups, with Renor, Galena, and I taking the west flank. "Do not allow them past you, but drive them toward the center. We will meet there. Keep your faces down and your voices low. Use gestures when possible."

Ren took point, and I put Galena between us so I could watch the rear. The ridges provided protection for a few scant meters, then began sloping down and playing out as ice hills and mounds fringing the broadly fractured surface of the plateau. Plas-Face made a complicated gesture, and Galena leaned forward to whisper an interpretation.

"Ren wishes us to go one by one to the next rise. When I arrive, he will move to the next, and so on until we reach attack point." Her wings fluttered under her parka as she rubbed her arms. Cold made her little face look pinched. "My feathers have frozen. I am afraid I will slip or stumble again."

"You won't." I took a moment to rub my numb hands over the two bulges on her back, trying to restore some circulation for both of us. "Just keep crouched over, at an angle, and you won't fall on your face." I waited until Ren darted to the next snow mound, then let her go. "Now."

Galena rushed forward, vanishing into the blue, horizonless immensity of the ice. The only way I knew she'd made it was when she glanced back and I saw her white face appear against the snow. Then she nodded, and it was my turn to run.

The surface ice felt dense and slick under my footgear, like arena turf after a hard rain. Automatically I adjusted my center of balance, keeping my arms out and my blades tucked in my sleeves. We progressed in that one-two-three darting fashion until we were together and crouched behind a tiny bank, just west of the enemy lines.

"Akkabarran kill zones are here"—Ren touched the lower left part of his chest—"and the necks. Ready?"

I nodded, then glanced at Birdie, who was so cold she looked drowsy. At once I gently slapped her face with the tips of my fingers. "Wake up, sweetheart; we have to go and kill a bunch of people."

Some of the daze cleared from her eyes. "I . . . I am ready."

We transmuted our blades to raen form, then at a nod from Renor ran out single-file onto the ice. Kol and the others attacked at the same time from the east, and battle ensued.

I forgot that I was cold, that we were fighting simulations. All the noise and discomfort left my mind and I reached for that no-mind, no-self calm I'd found fighting Kol and Ren. It came over me like a warm wave, settling my nerves, making it easy to lift my sword and thrust it into a chest or slash open a throat.

The Akkabarrans fought silently, as we did, with only grunts and gasps of breath punctuating the wild fire of their rifles and the faint whistling of our blades against the wind.

I killed five before I realized we were being outflanked by a secondary line, and sidestepped to get back to back with Ren. "They're coming around!"

He made the same gesture Kol had in the quad—*Guard the House*—and pointed to Galena.

"She's fine." I'd already glanced at her a few times, checking, and Birdie was holding her own. "We have to circle back, drive them in."

He cut a simulated warrior open from neck to groin, then whipped his head to the side. "No way around."

There was, but I'd need his help. "This nudge thing you do, can you do it to more than one person?"

"I believe I can. I have never tried." He ducked as another sim hacked at him with its scythe; then Ren darted behind him and ran his sword through its back. "What do you want to do?"

"Birdie's going to fly me over the line, and you're going to help her." I waved to Galena, who jumped through an opening between two approaching sims to join us. She was bleeding from a cut on her face, but the drowsiness was gone.

"They are spreading out to the south." She panted the words.

"I know." I jerked off her tunic and checked her wings. "Stop flattening them. I need you to fly. Right now."

She looked stricken. "I cannot."

My hands shook as I rubbed the bloodless arches behind her shoulders. I knew she had been making practice jumps almost every day in the simulators, when she thought the rest of us weren't paying attention. I'd seen her stay aloft for a few minutes, at best.

"Either you carry me over to plug this line, or we'll lose the fight." I turned to Ren, who was holding off stray attackers with concentrated effort. "You weigh more than I do, right?"

"Yes, by twenty kilos at least."

I faced her. "It has to be me, sweetheart."

"Jory, I cannot—" She closed her eyes for a moment. "Very well. I will try."

She kicked off her footgear, ran back a short distance, then darted toward me. About ten yards away she jumped and stretched out her wings. I caught her feet as she sailed over my head, and prayed my weight wouldn't snap them off.

Something invisible gathered around me and pushed up.

We hovered barely ten feet above the surface for several moments, and I had to yank my legs up to keep from being pulled down by the sims. Then the invisible force pushed again, harder, and Galena soared upward.

We were airborne.

It took only another minute to outdistance the flanking secondary line, and then we dropped to the ice. I caught her as soon as my feet touched down and steadied her as she folded her wings and sagged.

"Here." I stood her upright and tore off my tunic and footgear. "Put these on before you cube."

Her gray face looked wanly up at me. "You will freeze."

"I'm bigger and fatter than you. It'll take longer." I turned to face the rushing blur of the warriors we'd outsmarted. "Put them on and get out your sword; here they come."

We spent the next frantic minutes driving the enemy line back toward the center of the ice. Kol and his team managed to herd their side in as well, and we ended up facing each other with a dozen wounded Akkabarrans snarling between us. They were the most complex simulated warriors I'd ever faced, programmed to execute sophisticated attacks, and it took a while to kill all of them.

By the time we were done, I was covered from chin to knees with a heavy frost of frozen blood and gore, and so numb I couldn't feel my sword anymore.

I looked through gritty eyes at the others, who had a variety of small wounds and an equal amount of Akkabarran smeared all over their blades and parkas. "Are we having fun yet?"

A few seconds after the last body fell, the simulation vanished, leaving us pacing in a miserable circle in the center of the floor. The harsh yellow glare of the energized grid mesh faded as the trainer appeared, making notations on a datapad he carried.

He didn't even bother to glance at us. "You seven are exempt from training for the remainder of the day. Report to the infirmary for treatment."

"How about 'nice job, way to go, you nailed them'?" I planted my numb hands on my numb hips, and eyed the window in the simulator's wall. Uel still stood, watching. "Or is that too much to ask?"

Now the trainer swiveled to face me. "You do not deserve praise. This task should have been completed in half the time you took. Your attacks were slipshod and reckless. Prevailing winds should not have allowed your winged comrade to attain flight. And as for abandoning proper weather gear to protect another, on the surface of the actual planet, you would have died from exposure almost instantly."

Sparky threw out her arms, scattering everyone around her. "We endured this, only for you to tell us we failed? What was the point of the exercise?"

The trainer sighed and gestured to the door panel. "Report to medical. I will review the errors involved with this session in depth tomorrow."

I headed for the door, but my feet were so cold I couldn't feel the floor, and I stumbled. Nalek caught me before I landed on my backside, and swung me up in his arms.

Feeling clumsy and embarrassed, I pushed at his chest, but my arms refused to cooperate and went limp. In spite of my feeble lack of energy, I still protested with, "I can walk."

"You can fall, as well." He shifted my weight. "What you did for Galena was a noble thing, and I think you may have saved us all. Allow me to do this much in return."

Kol was right there, beside Nalek. And although he wasn't looking at either of us, I could feel how angry he was.

Yeah, well, you had your chance, pal. I let my head thunk against Nal's strong shoulder. "Okay."

As he carried me out, I saw the Blade Master leave the observation window.

The medical people patched us up, but they kept me in a berth overnight for observation. Kol got permission from Dursano to leave Nalek behind to watch over me, then came back a few hours later to relieve him.

"I'll live," I told him once we were alone. "Go get some sleep."

"How are your feet?"

"They hurt." I looked down the length of the berth at the bandages they'd wrapped me in from midcalf down. "They burn like they're on fire. Doesn't seem right, after being so cold."

"And your knee?"

I shook my head. "There's nothing wrong with it."

"You've been favoring it since we began blade training." He flagged down the doctor. "Healer, what of Sajora's right knee?"

"What of it?" The doc picked up my chart and switched it to display. "It's artificial, constructed and augmented with parts I've never seen used for an organic replacement, and the internal plates have deteriorated badly." He snapped off the chart. "I can't perform the kind of reconstructive work she needs."

"Find someone who can," Kol said.

"Let me handle this, Doc." When he left us, I struggled to get into a sitting position. "Listen, there's only one thing he can do about my leg, and that's amputate it from the knee down. However, it's my leg, I'm very fond of it, and I'm keeping it."

"I will speak to Uel." Unused to sitting, Kol carefully perched on the edge of a chair by my right side. "He can bring in a specialist."

"The guy who rebuilt this knee is serving at the front." I flopped back against the pillows. "I don't trust anyone else to touch it."

He gripped the handrail. "Then have them completely remove the cyber tech and replace it with what is normally used."

"Even if you could talk them into doing that, do you know what the recovery time would be? Six weeks, minimum." I snorted. "And in case you haven't noticed, there aren't any blade dancer rehabilitation units around here. If I can't train, Uel will kick me out."

Metal groaned under his fingers. "Is it worth your life?"

That was a pretty good question, and I considered it. My knee was only going to get worse. Every time I fought, I revealed more of my weaknesses to the other third-level trainees. They'd jump on the chance to disable me permanently. But I was more than one artificial joint, and I knew I could handle anything they threw at me. As for the knee, it had held together for eight years; it could stay together for a few more months.

"Yes." I looked at my hands, which were bruised and swollen from frostbite. "It's worth it."

He let go of the dented rail and took my hand in his. "Then I will do what I can to help."

"You could tell me something." I thought of what he'd said to me that night in the lavatory. "What, exactly, do jaspkerry and safira smell like?"

"Jaspkerry is sweet; safira is stronger, spicier."

That sounded like Chinese food, and made me smile. "Could be worse, I guess. You smell like rain on pine needles, you know."

"Needles?" He glanced at an instrument tray and made a face. "Surely not."

I chuckled. "No, they're leaves. Really thin leaves."

We sat in silence after that for a long time, and I dozed. The next time I woke up it was nearly morning, but it looked like Kol had been sitting up holding my hand all night.

"Hey." I sat up. "What time is it?"

"Nearly the hour to report for training." He paused. "I was angry at you yesterday. Your pardon, lady."

"None is required, warrior." I frowned. "You mean in session, or when Nal carried me out of there?"

"You did what was right, protecting Galena from the elements, but at great cost to yourself." He made a sound like a growl, and added, "I did not like seeing you helpless. You might have fallen with far greater wounds before the end of the simulation."

I lifted my shoulders, feeling uneasy. "You'd have done the same thing."

"Of course." His mouth curled on one side. "We are much alike in our regard for our kin."

"Oh, I don't know about that." I eyed Danea approaching through the ward entrance. "You like some of them a whole lot more than I do."

"She yet breathes." She stopped at the foot of the berth and propped her hands on her hips. "What does it take to kill you, Terran? *Two* armies?"

I returned the sneer. "Go raise them and let's find out."

"Thaw first, ambitious one." A few strands of her hair undulated and snapped before she turned to Kol. "Uel sent me with a message. You will return to training now. And you." She glanced at me. "If you do not return today, you will be recycled to second-level training."

That was all I needed to hear. I'd come too far to let a bad chill bounce me back to basics. I swung my legs over the side of the berth, winced, then began stripping off bandages. "Kol, get my clothes."

Maybe adrenaline was the only thing fueling my determination, but I dressed and accompanied Kol and Danea back to third level. When I entered bladework training, the others welcomed me, and everyone else gawked.

"How are you feeling?" I asked Birdie, then found myself in a light choke hold as she flung herself in my arms and hugged me.

"I flew," she said, and laughed, and danced around me fluttering before she hugged me again. "I really flew."

Over her shoulder, I saw Ren watching us, and needling guilt swept through me. She wouldn't have flown without his push, but there was no reason to tell her that now. Confidence was more important than honesty.

"You carried, too." I gave her a squeeze and set her down. "You are going to zoom rings around the other bird people in this place, you know."

She eyed some of the other winged trainees. "I think I may, ClanSister."

Bek showed up to teach the session, which surprised me. "Why do you stare, Saj?"

"No reason, Trainer, I just thought you'd be staying on second level."

"I go where I am directed. I also had similar expectations of you." He studied the seven of us. "You have done well in your first days. Strictly speaking, you should not have prevailed over the Akkabarran simulation, however"—his cheek twitched, and he cleared his throat—"yours was an innovative solution to resolving the secondary line."

I grinned. "So even the teachers are impressed? Damn."

He pretended not to hear that. "We will begin. Saj, you and Kol may demonstrate for the other trainees the proper positioning for pass attacks."

Pass attacks were the hardest and most complicated of long sword moves. Then I clued in on his wording. "The positioning, trainer?"

"Yes." The cheek twitched again. "Go through the patterns, but very slowly, so the class may attend you with ease."

Although my entire body ached, I'd been prepared to endure it and fight—not accept kindness and understanding from the same Chakaran who had been drilling us into the floor every day for the last several weeks. It was easy to be graceful and accept that. *Very* easy. "We'd be happy to."

As we headed to the exhibition platform, the door panel opened and two of the other trainers entered. Neither one looked happy. Bek consulted with them briefly, his own expression turning to stone before he left them to address the class.

"There is a change of schedule. All third-level trainees are to report to the challenge quad. This session is dismissed." He then came up to the platform, stopping me and Kol from following the others out. "Your injuries will not exempt you from participation in this, Sajora," he told me bluntly. "Stay paired with Kol; he can best assist you during the melee."

"Why are they doing this, Trainer? Is it some type of war games?" He nodded, and I took in a deep breath. No wonder the trainers looked worried. "I will, thank you, Bek."

"What is a melee?" Kol asked as we left the session room.

"It's an archaic battle term. My offcoach used it to describe our practice scrimmages. It means free-for-all," I said. "You take everyone you have, divide them into two teams, and have them fight each other."

"Uel cannot intend to do the same with the third level. There are too many of us."

We joined the others around the quad, where Dursano and other inductors were walking around handing out new bands—wide white bands with silver or brown triangles on them. Some trainees were handed one or the other; others were offered a choice.

When Dursano reached us, he said, "Do you side with the Faction"—he held the silver triangle band—"or the League?"

"Neither." Kol frowned, studying the bands. "As our homeworld has chosen, we remain neutral."

"Nevertheless." The inductor held out the bands. "For this practice you will select a side or one will be chosen for you."

With the exception of 'Gill, I had no fondness for the League, but I'd be damned if I'd end up siding with the lizards. "I'll take the brown."

"Wait." Kol took the band from me, reversed it to its plain white side, and wrapped it around his upper arm. "There is always neutrality in war, Dursano. It is its own side. You cannot deny us."

"You should take the brown." The inductor nodded toward a large group of trainees wearing silver bands. "They will force you to it."

Among the silvers were Fayne and a number of insectile, furry, and reptilian beings. As she tied her band around her throat, she looked up and gave Kol a beautiful leer as she waggled her fingers.

"Your girlfriend's siding with all the slaver species," I murmured. "Maybe he's right."

He shook his head. "Joren does not stand with the League. Neither shall we side with those who dishonor life."

As soon as everyone had been banded, the Blade Master mounted the platform and climbed with eerie grace into the quad, taking position in the center. Silence fell over the crowd as he was surrounded on either side by dimensional images of uniformed soldiers—a League colonel in dress brown, and a Hsktskt OverLord in thermal silver.

"The conflict between these two coalitions has entered a new stage. With each quadrant they invade, they either impress or recruit new species, and add to their already considerable forces. No neutral territory has survived an invasion." He turned full circle, zeroing in particularly on our little group. "You who do not ally yourselves may not have another opportunity to do so."

Despite that rather pointed warning, none of us took our bands off. Dursano and the other trainers began to separate the trainees, leaving us off to the side. Uel didn't let it go there, however. He left the quad and came directly to us.

"If you seek to avoid this exercise by choosing the white, reconsider. All of you will be pitted against both teams."

Sparky's hair bristled. "We are not cowards, Blade Master."

"Very well." He indicated the quad. "Prepare for engagement as a neutral force."

We entered the quad, followed by teams of silver and brown bands who outnumbered us two to one.

"This isn't fair," I said to Dursano, who indicated our positions from the floor. "It should be an even match, not fourteen on seven."

His thin lips quirked. "War is seldom fair, Saj, particularly to those who seek to escape it."

Everyone surrounding the quad began talking and laughing as they saw how we were positioned in the center of the opposing teams. Some threw out ribald suggestions on how we should fight.

"Give them your backs; make it quick!"

"Jump over them!"

"Kneel and beg; the browns will be merciful!"

Kol positioned us so that we faced out to both sides, with Ren and Danea in flank positions.

"Do not wait; do not give them the advantage," he said in a low voice. "Use the no-blade attack."

"The *what?*" I muttered back, and he put his mouth next to my ear and told me. At first it didn't process; then I realized what he was doing and laughed. "You're absolutely insane."

"It will work." He lifted his face to check the position of the hoverdrone. "On three—one, two, *three*."

Before the drone could finish opening the match, all seven of us attacked. Kol, Osrea, and I lunged at the silvers, while Nalek and Galena went after the browns. Danea and Renor went immediately to the sides, waiting on flank.

I yelled as I slashed out with my raen-tån, but I didn't bother hitting either of the two silvers in front of me. Instead, I followed Kol's instructions and cut the ryata cords behind them. The three elastic cords, which normally kept fighters in the quad, collapsed, leaving that side of the platform completely open. Using my arms and momentum, I shoved them both over the edge. They fell off the platform and hit the floor as the watching trainees scattered. Then I wheeled around and did the same to another silver.

The others also cut the cords and shoved their opponents out of the quad. In less than five seconds, ten challengers had been thrown to the floor.

Renor and Danea closed in on the four left, and made short work of them. Sparky used her field to force them over the side, and Ren didn't even have to touch them; he just looked and out they flew.

The seven of us resumed our positions in the center of the quad as Kol lifted his blade over his head.

"White neutral prevails," he said, in the stunned silence.

I made a point of catching Fayne's eye, and waggled my fingers at her. *Now who's laughing, Blondie?*

A moment later dozens of voices began shouting out in protest, and a couple of the embarrassed challengers tried to climb back into the quad. Dursano and the other inductors pulled them away as the Blade Master mounted the platform.

"The rule stands: Any challenger forced out of the quad loses the bout." His statement made the trainees fall silent again. "The ClanJoren has prevailed."

ClanJoren? Mom had been called that. It was a title, an award. They gave it to someone who sacrificed a great deal or did something important for the planet. It was also a way to make someone an honorary member of every HouseClan. Either Uel didn't know that, or he was being majorly sarcastic.

"We are not ClanJoren," Danea said.

Kol lowered his blade. "Perhaps here, we will be."

CHAPTER FIFTEEN

"THE TRAVELER AND THE PATH ARE ALONE UNTIL THEY JOURNEY TOGETHER."

—TAREK VARENA, CLANJOREN

From that point on, quad group exercises were integrated into our daily training schedule, and a new mood settled over the third level. Before, there had been plenty of rivalry and competition, but on an individual basis. Now it didn't matter what rank someone wore; everyone had become either a brown or a silver, League or Hsktskt, colonizer or slaver. Most of the trainers picked sides as well.

We heard wild rumors about the war, and the possibility that Reytalon might be evacuated. Someone said the staff had shut down the now-empty first level completely. More second-level trainees were advanced to third every day. We began seeing more reds and even a few orange bands among the ranks.

So many challenges were made by opposing members of both sides that the quad was never empty, and the bouts went beyond vicious. Sometimes as many as ten trainees died as a result of implant kills each day.

The entire Tåna had gone to war and, as we had in the very first bout, the seven of us landed right in the middle of it.

"I can deal with the jeers and the shoves and the challenges thrown in my face every five seconds," I told Kol one night in our quarters. "But I have a real problem when it's the trainers who are doing it."

"They are no more immune to this conflict than the others." He checked his blades and cleaned some sim blood off the blade guards before sheathing them. "We cannot abandon our neutral status now; some of the others are beginning to see the wisdom of it."

Nalek flopped down on his mat. "Several browns spoke to me about joining us."

"Some of the silvers are interested, too." Osrea pulled off his tunic and flexed his pinned arms with a grunt of relief, making his blue-plated hide crack at the seams. He caught my look and made a rude gesture. "Not all reptilians are interested in slave trading, ClanSister."

"I do not like it." Danea paced around the room with restless energy. "They have already cut our rations in half. When I spoke to the server drone, it claimed it had been programmed to withhold food from us. That neutrals must suffer deprivation during war."

"It is another tactic to force us to choose sides," Kol told her. "I will speak to Bek; he seems sympathetic to us. Perhaps he can intervene for us with the Blade Master."

"I doubt it," I said. "Uel is probably the one who gave the order to starve us."

Galena tugged me off to one side. "Is that why you gave me your meal tonight, Jory?"

I shrugged, embarrassed and ready to clock the snake-haired harpy for opening her big mouth about rations. "I really wasn't hungry."

She pressed a thin hand to my cheek. "Never have I had such a ClanSister."

Os, who had drifted over by us on the pretext of getting a clean tunic, gave me a very strange look—as if he wanted to hug me and punch me at the same time.

I checked the chronometer. "Kol, if you want to take that extra session with me tonight, we'd better go."

Since my injuries from the Akkabarran simulation had healed, Kol and I had by unspoken agreement put aside our personal problems to spend an additional hour after regular training to fine-tune our moves. Bek had instructed us to use an empty storage chamber just off the main bladework room, where we could secure the door panel and practice in peace.

We walked down the corridor, passing a few stragglers from the evening meal interval. A few of them had removed their obek-las and unfastened their tunics.

"Have you noticed the change in temperature?" I asked Kol.

"It seems warmer each day." He turned the corner, and stepped to one side to avoid a pair of disgruntled furred beings arguing in low growls. "You think it is intentional?"

"I think the silvers are mostly cold-blooded and are enjoying the hell out of it." I went to the chamber panel and entered the code Bek had given us. "You know they do better in hot climes."

"While tempers do not." He thought it over. "Why would the Blade Master seek to deliberately aggravate the trainee population?"

"The same reason he started these war games, though what it is beats me." Lately Uel had been present at every session I attended, too, although I hadn't mentioned that to anyone. I still wasn't sure he was there to watch me, and none of the trainers would comment on his presence. "Where did we leave off last night?" I asked as we entered the room. "The running attack?"

He scanned the room, then drew out his tåns. "We should concentrate on your remise. You need to trim more reaction time from your parry to counterattack."

I pulled my blades and rolled my shoulders, trying to loosen up. Already sweat was beading above my brows and upper lip. "I can do that."

We faced each other, tapped right blades, and began to spar. Advancing until there was no more room to occupy, we avoided each other's initial feints and parried the simultaneous, genuine thrusts.

"You are aggressive tonight," Kol murmured, probably in an attempt to distract me.

"Yeah, I do feel like kicking your ass again."

Our tåns collided, parted, met again as we began circling each other. I fell into the rhythm and lost sense of myself, knowing only to where to move and when to counter Kol's blades. The heat was a problem, though. It felt like someone had sucked all the air out of the room, and sweat was making it hard to focus.

"Time." I stepped back and held out both hands, blades up. As he lowered his blades, I saw he was dripping wet, too. "It's getting too hot for this."

He pulled off his tunic, revealing his wide, sweaty chest. "I will see if I can adjust the envirocontrols."

While he worked on the console, I stripped out of my tunic and trousers. The one-piece undergarment I wore was soaked through, but I felt better.

"I can't override the shutoff program," Kol said as he came back over to me. "We must continue tomorrow."

"Why?" I dried my face and hair with my tunic, then saw how politely he was averting his gaze. "Oh, Christ, Kol, you've seen me naked a couple times. Take off those pants and let's go."

"It is not proper to fight unclothed."

"I hope you never get attacked while you're cleansing." I grinned. He was such a Jorenian for formality. "Come on, it doesn't matter. Let's spar another round; then we'll call it quits."

With visible reluctance, he stepped out of his trousers and kicked them aside, then faced me and held out his right blade. "Engage."

I tapped, then forced his blade to one side, stepping in fast to meet his left blade, coming up toward my abdomen. "You're going to have to be faster than that, pal."

He pushed me away and circled to my left. "I am."

I hooked his right blade and clashed mine with his left, crossing our arms until I disengaged one and he the other. "Nice. Try again."

We sparred with renewed speed, and I enjoyed the immediate benefit of shedding my garments as I glided around him unimpeded by fabric. The sweat on our skins made us both slippery, which led to us both nicking each other a few times until we compensated for that.

Slipping into the no-mind rhythm seemed more difficult now, though. With our bodies exposed, I was acutely aware of him, of the slickness of his skin as we locked arms and blades.

His scent blended with my own, and became something darker, more intense. The play of his muscles distracted me, too—he had toughened up even more over the last months, and his long limbs extended and bunched with sheer animal grace.

My damp feet slid on the floor, and one of his blades nicked my shoulder. I rolled around most of the thrust and tried to return it, only to lock blades with him again.

An inch from my face, he asked, panting, "Enough?"

"No." I applied pressure to his wrist, knowing I had the better angle. "What, are you tired, ClanBoy?"

He wrenched out of the lock and slashed at my hip, but I spun away before he could cut me. When I looked down, I saw the shoulder hit had caused half of my top to peel down, and thin rivulets of blood trickled over my exposed breast.

"I ought to make you mend this," I said, spinning my blades and stepping to his right, egging him on. "I hate to sew."

"As do I." He lunged, going for an opening I'd left on my right, then looked down to see my blade cutting into the waistband of his shorts. He whirled out, trying to get behind me. "Payment in kind, Sajora?"

"Eye for an eye. It's a Terran thing." I countered the move. "Not getting modest on me again, are you?"

"No." He came at me again.

The dance changed. We were slashing at each other, forgetting form and defense as we plied our blades toward the win. Kol struck again at my shoulder, sending the other half of my top drooping down, but I slid the tip of my blade down his hip, completing the cut I'd made and leaving his shorts hanging from one buttock and leg.

Breathless but feeling confident I'd win—my undergarment was tighter, and therefore harder to cut off than his—I pivoted, going for a side pass so I could swipe at his backside from behind.

My knee gave out.

"Shit!" I yelled, wildly throwing out a hand. His left blade slashed across my palm before it dematerialized, opening a gash from fingers to wrist. Then I went down, hard. "God *damn* it."

"Sajora." He let his blades fall to the floor as he dropped to his knees, hoisting me up, cradling my hand in his. "Your pardon. Mother of All Houses, I did not mean to harm you."

"You didn't. My knee blew." I bent over, trying to see the damage, and smacked into his chest. "Would you move—"

"Be still." His hand caught my chin and forced it up. Heat blazed in his white eyes; then they blurred as his mouth covered mine.

One vague part of my brain reminded me that we weren't supposed to be doing this. Then it shut down as I sank into the kiss. My wet fingers ran up the tense muscles of his arm, smearing his skin with my blood before I tangled them in his short black hair.

Without lifting his mouth he knocked me over, coming down full weight on top of me, pinning me to the floor. I curled my legs around his, adjusting my hips until I had his erection pressing against where I wanted it, where I needed him. When his hand moved down and cupped my breast, I pushed up against his fingers.

He tore his mouth away. "Sajora—"

"Kol." I bit at his mouth, catching his lip. "Don't."

"I cannot do this." His white eyes were mere slits, his face as tightly set as his body against mine.

I knew the Jorenians had a lot of formalities about relationships, but we weren't getting married, and I had no intention of letting him go without both of us getting some basic satisfaction.

"You can't do this?" I kissed him, hard, then let go of his hair and reached down between us. With a little tugging I pushed the wet fabric aside so he could feel the folds of my sex nestled against the hard dome of his shaft. "Or this?"

"It is wrong." He rested his brow against mine, his teeth clenched, his entire body still. "Stop this; I cannot think."

I shifted my hips so that I rubbed against him. "You're about to stop thinking altogether." I brushed my lips over his cheek, kissed his ear, then rolled until I ended up on top, where I had control of what went where. "Let me demonstrate."

"Sexual activity is confined to personal quarters."

We both froze as the Blade Master materialized right beside

us. I felt Kol's arms tighten around me, and I sighed. "Great timing."

Kol lifted me off him and got to his feet, hauling me up at the same time. "Uel." He handed me my tunic and trousers, then shielded me from the Blade Master's gaze with his body as he pulled his on. "Sajora and I will leave."

"Go to the infirmary first and have those wounds checked."

I dressed, trying to resist the urge to take out my blade and stab Uel in the heart. As we moved past him, the Blade Master held up a glove. "I will speak to you alone, Sajora."

Kol hesitated, but I waved him out. "I'll catch up in a minute." When he was gone, I turned to Uel. "Well?"

"Dancers should not take mates."

The Blade Master was giving me advice on my love life. It was too bizarre for words. "And?"

"If you wish to work off your sexual frustration, I would suggest you choose a partner outside your clan."

"Uh-huh." I tilted my head to one side. "Like you, maybe?"

He stepped away from me as if I'd suggested something gross. "I do not wish to have relations with you."

What was it about me and blade dancers, that none of them wanted me? A girl could get a real complex from that kind of steady rejection.

"So you've been watching me every session for the last couple of weeks purely out of aesthetic interest?" I didn't wait for a response; it was enough to let him know I knew. "I don't need or want your advice, Blade Master, but I'll let you know if that changes."

"You and Varena are mirrors of each other," he said. "If he becomes emotionally involved with you, it will destroy him."

The grim prediction made me grow cold. "You've got me mixed up with Fayne."

He made a sound that might have been a laugh. "Fayne is a killer without emotion. You are the daughter of one."

I zeroed in on the last part. "You know Kieran that well?"

"I know him by watching you." He disappeared through a hole that opened, then closed in the wall.

"Wait!" I ran over and tried to find the passage so I could go after him, but there were no seams, and no panel to access. Furious, I punched the alloy, then went to find Kol.

I caught up with him just outside medical, but he didn't say a word to me. I was too frustrated and ticked off at Uel to come up with idle conversation as we went in and were examined by the doc. His cuts were all minor, but my hand required sutures.

He stood by the berth as the doctor stitched me up, waiting with the heat still simmering in his eyes.

I wasn't bad for him. I could take the heat, give it back—and then some. We only needed to clear up the matter of whether he was Kieran's son or not.

"Doc, what does it take to do a DNA comparative scan?" I asked as he finished bandaging my palm.

"I can perform the test in a few minutes. Why?"

"Excuse us for a minute, will you?" When he left, I gazed steadily at Kol. "You heard the man. A little blood, a few minutes; then we'll know."

"You mean this test will reveal if we are siblings." I nodded. "No, Sajora. I will not take it."

I should have shouted at him, but the look on his face made a knot tighten in my chest. "I need you to do this for me, Kol. I have to know now. After what almost just happened, I think it's an imperative."

"We are kin." He reached out and brushed his fingertips over the short, curly hair that had grown back on my scalp. "As to whether we are blood kin, it matters not."

"How can you say that?" Despite my irritation, all the frustration seemed to drain away inside me. That's what he did to me with a single touch. I wanted to turn so I could press my face against his palm, but I couldn't look away from his eyes. "Please do this. For my own peace of mind."

"If we are blood, then what we feel must perish." His finger sketched a delicate trail over my brow, down my cheek. "If we are not, then what we feel will perish."

"If we're not, it doesn't have to."

"You do not wish to bond with me, and I will not force you."

Fayne is a killer without emotion. You are the daughter of one.

"Okay. You need us to do this Choice thing first; I can do that," I told him. To hell with what Uel thought. I touched his face. "If I've got to marry you so we can be together, I'll do it."

"What of your vow?"

My fingertips went still. "I haven't taken any that I'm aware of."

"You said that you would never have children."

"Kids?" I frowned. "Yeah, well, I don't want any. I'd be a terrible mother."

"I am Jorenian, not Terran. Children are why we Choose, Sajora. To bring new life to the path."

It was my own fault—I'd been very straightforward about my dislike of children. I didn't want to be a brood mare; I had better things to do. And what sort of mother would I have made, anyway? Parenthood simply wasn't an option.

It wasn't an option, and still I found myself thinking about it. Having Kol's child growing inside me, getting heavy, waddling, writhing as I tried to force it from my body. I looked into his eyes and swallowed hard.

"I guess I could try it and have one," I said at last. "I mean, everyone has kids, right? I'd probably get used to having it around after a while."

He pressed his fingertip against my lips. "You would bear my child, and it would either be deformed because we are blood kin, or unwanted, like us."

"*You* were unwanted," I shot back. "*My* mother gave up her family and her friends and her whole life for me."

"Perhaps the shame was too much for her, and she ran away." He grabbed my fist before it connected with his face. "Why else would she select a place where she would have to hide from everyone for the rest of her life?"

"She wanted me to be free, instead of becoming an unpaid servant, like the rest of you."

"She wanted you to be Terran, and she succeeded. But why, Sajora?" He forced my arm down, but held on to my wrist. "Kalea Raska was a strong woman—intelligent, gifted, and much loved among my people. Why would she run away to raise her only child on the homeworld of the man who had dishonored her so?"

"I told you." I tugged my hand free and rubbed my wrist. "Her HouseClan kicked her out; it's their fault."

"What are you hiding from me?"

I refused to answer him.

"Very well, my heart. You may keep your secrets. But know this." He bent over and brushed his lips over my brow. "Had she stayed, you would believe in Choice as I do. As all our people do. Kalea denied you that."

Out he went.

Something changed in me after that gentle but cruel rejection. I felt almost the way I had right after Mom died—I didn't want to eat, sleep, or be sociable. Kol must have said something to the others, because everyone tiptoed around me, and even Sparky stayed out of my face.

Kol spent more time training outside the clan, but I pretended not to notice. I didn't talk to him, and I stayed as far away from him in any situation as possible. It wasn't because I was angry at him. Blaming Kol for making me face an unpleasant truth about myself would have been unfair, so I told myself to let it go—and I thought I had, until one particular sparring match.

Bek had me warm up with Dag, one of the wide-bodied humanoids who had shoved me around over on second level a few times during Hell Week. He acted like he'd completely forgotten about it, but I hadn't. That's why I wasn't surprised by what he said when we took our positions.

"Fayne sends her affections," he told me, flashing his silver band. "She's offered a bounty for anyone who can cut up your face."

"Still won't challenge me herself, huh? The little skink." At

Bek's signal, I rolled out of Dag's attack line, ducked under his arm, and came up behind him. I pinned his arm, twisting the end of it up behind his back. With my other hand I sank a blade into his chest. "Did Blondie mention what I'd get if I cut up yours?"

I broke away as he wrenched free and spun, then slashed my blade across the back of one of his hands, making contact long enough before the blade went holo to slice through half a dozen tendons and render his grip useless. One of his tåns fell to the floor, drawing the attention of the other trainees.

I ignored the hoots as I circled him, sizing up his guard, which was sloppy and full of holes. I could have stabbed him in the chest again at least a dozen times, but I wanted to play with him for a bit longer.

"I'll hack out your eyes and put them on a cord for her." His blade whistled past my face, but the tip missed my cheek by a hair. "She can wear them around her neck."

I brought up my elbow, knocking his chin up and exposing his throat, which I also slashed. The blade didn't go deep enough to nick any arteries, but I liked seeing his flesh part and his blood drip down the front of his tunic. "Worry about your own throat."

"Saj," I heard Bek say behind me. "You are to spar, not toy with him."

That seemed to infuriate Dag into making a wild lunge at me. Overextended as he was, it was simple to trip him and knock the last blade out of his hand. I planted a foot on his chest, bent down, and stabbed him in the implant a second time.

"Don't kill me!" He looked frantically around for help. "Please, I concede."

I smiled down at him and raised my blade again. Light glittered on my claws. "Say pretty, pretty please with sugar on top."

"Sajora." Bek hovered at my side. "This is a sparring match, not a challenge."

"It is." Dag nodded with terrified little jerks of his head. "We were only supposed to practice moves."

"This piece of shit promised to carve my eyes out, Trainer." I sheathed my tån, then brought one of my claws down until the very tip of it hung a scant millimeter above his right eyeball. "I figure he owes me one just for making the threat."

"You have prevailed. Let him go."

I traced a tiny, invisible circle above his eyes. "I want him to say I prevail. Loud enough for everyone to hear it."

He gulped, unable to blink without slicing his eyelid open on my claw; then he shouted, "You prevail; you prevail!"

"Good boy." I stepped off him and watched as he scrambled to his feet.

"Dag, report to medical." The trainer gestured for me to accompany him out of the room as well.

Nalek tried to say something as I strode past him, but I wasn't interested in more clan nagging.

"I didn't skip any of the pivots," I said to Bek as we made a circuit around the quad.

"No, you did not." He looked up at me. "You heal very fast, for a Terran."

I regarded my claws, which oddly had not yet retracted. "Blame my bad Jorenian blood."

"In there." Bek stopped me by pointing to one of the empty planning rooms the trainers used. When the door closed, he rubbed a paw over his scarred face. "I should send you back to second level for what you did to Dag, but we are closing that section soon."

I sat down on a table and held on to the edge. It was just high enough for me to swing a leg back and forth. "Do what you have to do, Bek. I've never asked you for any special favors."

"Not even when you hear that I trained your father?"

I jumped off the table. "You did what?"

"You've been asking everyone about Kieran," he said. "Perhaps it is time you got some answers."

I hadn't told anyone Kieran was my father. All Uel knew was that I wanted him to be my first kill as a dancer. "Who told you that Kieran is my father, Trainer?"

"No one, Saj." He smiled a little. "You are unmistakably his offspring."

Was I. "Do you know where he is?"

Bek shook his head.

I didn't even realize I was going to ask the next question until it popped out of my mouth. "What is he like?"

"He's efficient. He learned the no-self discipline before he ever came here." Bek's two remaining whiskers twitched. "He never lost a challenge."

I waited, but he didn't add anything else. "And that's it? He's a good fighter who never lost? That's all?"

"I would happily bring Kieran here and put you both in the quad to settle this with blades." Bek uttered a terse laugh. "I might even tell him that you're his daughter, after he finished chopping you to pieces."

That pissed me off. "You never know, Trainer. I might win, considering."

"Kieran is the most natural fighter I ever trained. No one could touch him, not even your Kol." He gave me the once-over. "You have his blood in your veins, but it is tainted with the foolish emotions of an overwrought female."

I could have stabbed him for that. "This overwrought female just whipped your big bad silver back in the session room."

"I pitted you against Dag because he is slow to react, overconfident, and otherwise stupid. Had you sparred with anyone else, they might have given Fayne a gift of your eyes." The Chakaran shook a claw at me. "Leave your petty concerns outside my training room from now on, or I will instruct them to keep second level open so you may dance with the drones."

"Fine. Will you help find Kieran when I take the black band?"

His pupils dilated. "I cannot leave the Tåna, but I will do what I can."

"Wait a minute—you can't leave? Why not?"

"That is order business, trainee." Bek jerked his head toward the training room. "Get back to work."

Along with pleasing my disgruntled trainer, I had to face new challengers in the quad each day. After the disaster of the first three-way melee match in which we'd shoved the silvers and browns out of the quad, we were never again matched as large groups, but neither did we fight alone. Most of the time we were either paired off or sent in a threesome to fight.

And we weren't alone.

Other trainees began turning their bands from brown and silver to white neutral, and fighting beside us. More browns turned than silvers, but that was to be expected. Most of the cold-blooded were either friends of Fayne's or terrified of her. Despite that, after a week the white neutral group had tripled in size.

"In the war, my people also remain neutral," a Tingalean named Mojag told me after we'd won another match fighting together. "It is not a position of great ease."

"Tell me about it." I rolled an aching shoulder. "Your moves are good; I'd never have guessed you have ten other limbs."

"We are taught from hatching to use each pair independently." Mojag nodded toward Osrea. "Your blue-plated comrade there was not."

That gave me an idea. "How are your kind taught to do that, exactly?"

What he told me made a lot of sense. I asked him to meet me and Os after training, then stole some fruit out of the dining hall, hiding it under my clothes. Later, I cornered Osrea in our quarters, told him what I wanted to try, and dragged him out to a practice room.

"This will never work."

"Give it a try." Once Mojag entered, I secured the door panel and took out six small near-apples. "Mojag, this is Osrea. Os, take off your shirt."

"He has twelve limbs and a tail; I have six." Still grumbling, he stripped, releasing his other arms. "And I do not walk on my abdomen."

"Neither do I," the Tingalean said. "I slither." He shrugged out of his shirt and extended the short limbs on either side of his snakelike body. "I am ready, Saj."

"Okay, Os, watch." I started pitching the fruit at Mojag, who caught the first two with his upper pair of limbs. I deliberately threw the second two at his lower limbs. Instead of catching them with another limb pair, he tossed the first two near-apples into the air, bent, and caught the second two. By the time I'd thrown the last two fruits, Mojag was keeping four in the air, and had used only his two uppermost limbs.

The other ten didn't move because the Tingalean had clasped them together.

Osrea watched, then hesitantly joined his lower arms together the same way. "Like this?"

"That is only part of the exercise, *warrus*," Mojag caught four of the near-apples with his upper limbs and the last two in his hinged mouth. He jerked his jaw up, tossing them back to me. "When you have learned control, you will no longer need to hold yourself thus."

"What does *warrus* mean?"

"End hatched." The Tingalean slithered over and held out an upper limb. "My kind, we look out for the last from the nest."

I imagined a little Os eating his way out of a shell and bit the inside of my cheek.

Snake Boy still looked suspicious. "You would teach me to do this, as you do, with your limbs?"

"Of course." Mojag arched his neck, making two curved folds of skin stand erect. I immediately thought of a Terran cobra getting pissed off, but the Tingalean was giving off different vibes. He was being generous, or friendly maybe. What he said next confirmed it. "We share nest-blood, little brother, your kind and mine. It is my duty and my wish."

They soon forgot about me as Mojag showed Os how to move and flex his limbs to gain more control. My clumsy ClanBrother began to move differently, not as a Jorenian but as a Tingalean would. It made all the difference, and I could see that in no time Os would attain the same lethal grace of movement that Mojag possessed.

"I would like to practice with you again," Os said to the Tingalean as we left the session chamber an hour later. "If you are willing, *tevhat*."

Mojag had told my ClanBrother to call him that, and that it

meant "soil breaker," or the first hatched from the nest. "That is my wish, too, little brother."

"I'd like to watch more of your moves myself," I said. "If you don't mind an audience."

"Like to like, and unlike, yes. I think we have much to learn from each other." The Tingalean nodded his wedge-shaped head. "So we shall."

When we entered the corridor leading to our quarters, I spotted Kol standing in an open doorway, and frowned. He was bending over and speaking quietly to someone inside the room.

"Excuse me." I left Mojag and Os and went to see what was going on. A few feet from the door I saw a slim white hand resting on Kol's hip. No, not resting—moving in a slow circle, the fingers caressing.

My blades were in my hands before I realized it. "Kol?"

He turned his head toward me for a moment, then said something and stepped out of the doorway. "Go back to our quarters, Sajora."

Fayne stepped out of the room. An open robe of some transparent pink stuff floated around her dinky body. Her dinky *naked* body.

Sexual activity is confined to personal quarters.

"What's going on?" I asked, my voice turning ugly.

"Ah, the Terran with the limited vocabulary." Fayne slid an arm around Kol's waist like she had a right to put her hands on him. "Perhaps she would like to join us."

"No, but I have no problem with separating your head from your neck." I shifted my gaze to Kol. "Her? You can't be serious."

"You are intruding." He kept his expression blank, but flashed me the Guard the House gesture. "Leave us."

As much as I wanted to gut the naked little midget, I pivoted and headed for our quarters. And waited there, watching everyone sleep and staying on guard, for the rest of the night.

Kol never returned.

CHAPTER SIXTEEN

"Divert a path without cause and condemn your soul to eternity of wandering."

—Tarek Varena, ClanJoren

I left our quarters early the next morning, and went into the third level alone. The idea of pounding something to a pulp appealed to me.

A lot.

The moment I emerged from the corridor, I saw Fayne and Cirilo hanging out at the quad with a group of silvers. An equal number of browns occupied the other side of the arena, and a few whites were segregated off to one side.

Everyone looked at me, then Fayne.

The midget and her merry band began to move in my direction. There were no trainers out yet, and the hoverdrones couldn't do anything but squawk at us if the silvers attacked.

I could deal with that.

The whites came over to flank me. "Where is the rest of your clan?" one of them asked me.

"Sleeping in." I drew out my blades.

"That is not good."

"That's the way it is." I met Fayne's colorless gaze across the quad. "You can take on anyone you like, but the midget albino rat is mine."

Another white hissed in a breath as the silvers armed themselves. "You're welcome to her."

"You rise early, clod." Fayne stopped a few meters away and examined my group. Her blades gleamed in her hands. "Even with your backup, you're outnumbered—or can't you count?"

"Hasn't been a problem so far. Kind of like you." I smiled. "Sleep well?"

"Hardly an hour." She trailed white fingers down the front of her shirt, stroking her own breast. "Kol is a magnificent male, isn't he? So strong and full of stamina. A female could spend years pleasuring herself with him, and never begin to tap his potential." She let her secondary eyelids droop. "I may even take him as a mate, when we join the order."

My claws emerged. "You're not going to make it that far, Blondie."

She bared her pointed little teeth. "Do you mean to challenge me? After all these weeks of running away at the sight of me?" Her cronies made various sounds of amusement. "Why, clod, you've finally discovered your spine. We should announce it. Perhaps the Master will declare a holiday."

One of the whites standing next to me clamped claws around my forearm. "Saj, she will kill you."

"She can stand on her toes and try." I shifted my grip on my blades. "You want to dance, Blondie? Let's—"

"I am pleased to see so many of you here at this hour." Bek strolled between the two groups. "Fayne, you and the silvers may begin sparring in the bladework room." When she didn't move, he folded his arms. *"Now."*

"I have a challenge to fight first," she said. "This Terran practically begs for it."

The Chakaran eyed me. "Did you challenge the Skogaq, Saj?"

"No." Kol joined Bek. There were faint purple shadows under his eyes. Guess he didn't get any sleep, either. "There was no challenge offered, and none will be taken."

The pounding voice inside my head screamed in outrage, so loud that I could barely get a word out edgewise. "Shut up, ClanBoy, and stay out of my face."

My ClanBrother turned and slashed a blade at my face. I reacted so fast that I knocked over one of the whites, but the tip caught my right cheekbone. I felt the sting, then the warmth of blood trickling my face, and watched with wide eyes as Kol resheathed his blade.

Cut me. He cut me.

I couldn't move, couldn't breathe. All I could do was touch the cut on my face and stare at him.

"Sajora is injured," Kol told Bek. "She cannot challenge, or accept a challenge, until her injury is treated."

The Chakaran nodded his scarred head. "Report to medical, Saj."

In the end, Bek ordered two whites to escort—well, *drag*—me to medical. The doc threatened to put me in restraints if I didn't sit still for the sutures, so I shoved back my temper for a few more minutes. I passed on the anesthetic and sat under the burn of the laser, letting the pain feed the beast inside me.

"You will be scarred," the doctor told me.

"Good." Like the spit-bath when I was deported, I needed to remember this.

I went back to third level alone, my face throbbing, my blades in my hands. Bek ordered me into the quad and sent three browns in after me. They didn't stay long.

I didn't let the beast go, but I gave it a lot of chain to play with.

An hour later I stood covered in sweat in the center of the quad, with two silvers and a brown flat on their backs before me. The crowd of trainees who had been watching the match had put a foot of space between themselves and the ryata. Even the two whites Bek had sent in to fight beside me for the third challenge kept a safe distance between themselves and my blades.

So I was feeling a little aggressive. My face hurt. People were just going to have to deal with it.

"Bek." The sound of my voice sent one of the silvers scrambling out of the quad in a panic. "Send in the next batch."

"You have already defeated nine opponents; I believe that is adequate for this session." The trainer nodded to my teammates, then held out a towel as I climbed down. "Your face is bleeding again. Perhaps you should return to medical."

I wiped off the sweat and blood, and wondered if he knew I was pretending every opponent he threw at me was Fayne. Or Kol. Emotionally overwrought female that I was.

"Os appears to have found his balance." Bek nodded toward Snake Boy, who was engrossed in a conversation with Mojag.

"He's given up trying to be a ClanBoy and is getting in touch with his snake side." Automatically I looked for Galena, and saw Sparky and Nal standing guard. "Now if I could get the rest of them . . ." I trailed off and turned to scan the assembled white neutrals. They were all assorted sizes and species. Familiar sizes and species.

Like to like.

The tidal wave of rage churning inside me finally subsided a few inches. "Excuse me, Trainer."

The avatars kept their wings bound during training sessions, unless in the simulator or common areas. The two I approached fanned their primary remiges, a show of avian readiness for flight or to fight.

"Hey." I held out my empty hands, palm up. "We're all wearing white neutral here, relax. Strix and Tej, right?"

Strix bobbed his feathered head. "We thought you'd gone color-blind today, Saj."

"Yeah, well, I'm in a bad mood. Never spar with me when I'm in a bad mood. You see my ClanSister over there?" I nodded toward Birdie. "She's got a problem with her wings."

"We have noticed," Tej said. "She does not use them."

"That's not her fault. She was raised on Joren, with ground dwellers. They didn't teach her to fly, and they kept her plucked." I measured the way their wings arched. "I've been coaching her, but she needs tutoring from her own blood. Would that be one of you?"

"No. We are Suruki. Your sibling resembles the Calypte." Strix turned to Tej. "You fostered with her kind as a fledge, did you not?"

"Two seasons, until my foster dam shoved me out of the *weyre.*" The avian studied Galena for a moment. "She is *lustrial,* not *dyratane.* A carrier, not a predator," he added at my blank look. "We are *dyratane.* We fly to hunt, not to collect nectar and pollenate flora."

"Blade dancers don't have a lot to do with flowers, do

they?" Strix and Tej exchanged a glance; then Tej shook his head. "My ClanSister can fly, but she needs to learn how to survive. The things she needs to learn on the ground, I can cover. On wing, I can't. You can."

A few minutes later I went over to tell Birdie about her new private dancing lessons.

"With him?" Her voice squeaked as she stared at Tej. "He's a Suruki. My ClanMother told me they eat raw flesh."

"You'll be training, not dining together." I patted her cheek. "And I'll be there with you the whole time, just in case."

Kol appeared, looking tired and grim. "Sajora, I would speak to you."

"Excuse me, Birdie." I turned and strode off for my next session.

"Sajora." He caught up with me. "Hear me."

"I heard you just fine the last time, you bastard." I sensed he wanted to grab me somewhere, and stopped. "Don't even think about it." I tried to walk away again. "Get away from me."

He paced me. "Fayne will kill you in the quad. I had to stop it somehow."

"Next time, cut her."

"She wishes to form an alliance."

I laughed once. "Right."

"I was obliged to listen to what she proposed."

"It took her six hours to tell you? What does she do, stutter?" I didn't wait for an answer, but entered the timing training room, pushing through other trainees to get to my spot.

"Sajora, this is important. There are other concerns now that we must attend to immediately." Kol assumed his position beside me as the trainer walked the rows. "I have much to relate about it."

I caressed the hilts of my blades, imagining both of them buried in Fayne's chest. Or Kol's. "I don't want to hear about how much fun you had fucking Blondie. Understand?"

"It was not like that." He sounded mortally offended.

"Oh, right, I forgot. You're supposed to be saving yourself for your wedding night." *With someone besides me. Fayne? Would he Choose the little skink over me? She wouldn't live very long if he did.* "You're giving abstinence a bad name, you know."

"Kol. Saj. Since you attend each other instead of me"—the trainer gestured to the center platform—"you may demonstrate your physical prowess in avoidance."

Since I'd had no sleep, and my eye was swelling shut, and I'd already expended most of my fury in the quad, I thought I'd be slow to react. But as the beams began flashing around me and Kol, I moved into the patterns of avoidance without thinking.

"There are rumors that the war has extended into this system," he said as he brushed past me to dodge two intersecting beams. "The Hsktskt and the League could be preparing to invade Reytalon. If they surround the Tåna, we will be trapped."

"Guess that wrecks the idea of taking her somewhere off-planet on your next date." I ducked and rolled my shoulder, feeling a beam so close the heat warmed my skin. "Maybe you should try a stroll on the surface. I'd get her to put on a little more than that pink outfit, though. She might catch cold."

"She attempted to initiate relations with me, however—"

I bumped into him deliberately, knocking him into a beam and enjoying the way he flinched as he took the jolt. "Didn't you hear me? I don't want to know. Go brag to the other boys."

"I did not touch her," he said, furious now.

"She not into virgins?" I dropped as a cluster of beams shot over my head. "Let me guess, she doesn't want kids either. Poor Kol. What's an honest Jorenian guy have to do to get laid around here?" I flicked a finger at my cheek. "Cutting a woman doesn't seem to work."

He grabbed me by the front of my shirt. "You will not speak to me thus."

I had a blade at his throat a heartbeat later. "You smell like Skogaq, ClanBoy. Don't get it on me."

The beams abruptly disappeared, and the trainer appeared beside us. "This session is focused on timing, not bladework. Put away your tån, Saj."

Neither of us moved. "We can take it out to the quad," I offered. "Want to go one-on-one? Make my day and say yes. Please."

"No," he said, his lips barely moving.

The trainer gave me a nudge. "If there is no challenge, there is no reason for this."

There were all kinds of reasons, but I saw Galena watching me with wide eyes, and reminded myself I still had five other reasons not to kill myself or Kol over Fayne.

"Fine." I backed off and walked out.

"Why did Kol cut your face?"

I had sat by myself intentionally to avoid clan interrogation, but Sparky never was particularly sensitive to someone's mood. I looked over my shoulder, and briefly imagined punching her in the face. "Go light up someone else's life."

"Answer me, or I shall make it yours."

"He didn't want me to fight his girlfriend. Now get lost."

She sat down beside me and leaned in, close enough to make the hair on my neck rise. "You are out of control."

"No, but I'm getting there." I dug my spoon back into my scanty portion of stew. When she didn't leave, I sighed. "Look, Sparky, would you just go and bug Nalek for a while? I'm not in the mood for another bitch-slap session."

"I have no desire to . . . bitch-slap . . . you." Danea looked at her footgear. "When Kol cut you, was there a threat to your life?"

I snorted. "Hardly. I was getting ready to fight the little midget, but he fucked that up."

Her hair did a little dance. "I thought as much." She muttered a particularly vile Jorenian obscenity.

"I totally agree."

"There is something you do not understand, and Kol is not thinking clearly." She bit her bottom lip and glanced back at Nalek. "Nal is correct in thinking it will get one or both of you killed."

"And this something is?" I made a rolling motion with my hand.

"Nalek believes you and Kol have warrior-bonded. After what I saw you do in the quad today, and Kol's attack on you, I must agree with him." She met my gaze, and instead of the usual nastiness, there was something like sympathy in her eyes.

"Sajora, if it is true, you will wish to kill anyone who violates your bond. As will Kol. It is extremely dangerous to go on in this fashion, both of you denying what we can all see before us."

I was so sick of the Jorenian bullshit. "And the reason I should care is?"

"You must declare the bond, and Choose Kol."

"You know, I tried to do that before," I said, with extreme patience. "He turned me down."

She drew back a little. "You Chose, and he denied?"

"No. I suggested we do the Choosing thing, so we could have gratuitous sex without guilt, and he refused."

Sparky thumped a fist down on the table. "Why would he say such, when it is so obvious that you have already bonded in every way but the word?"

"I don't want kids, and he does. I don't think you get any more basic a non-bond issue than that." I pushed my bowl away. "Satisfied now?"

"Wait." She looked even more uncomfortable. "I have never done this. It was for your ClanMother to tell you."

"Jesus, Sparky." I had to laugh. "Relax; I already know where babies come from."

"The Mother's blood runs through your veins, Sajora." She got all serious on me. "I never knew how strongly until now. You cannot deny this bond; you cannot fight it. It is the most sacred of things. Your soul Chose, and it will not be denied or thwarted."

"He's not so hot. My soul will get over it."

"That is not all. When you do not claim your Choice, as you are doing now, the chemistry in your body changes. You will grow more and more aggressive as time passes. As will Kol. In the end it will kill you both, and the only way to stop that is to declare your Choice."

"Dying from unrequited lust." I rolled my eyes. "Oh, sure, that'll be the way I go."

"How many times have your claws emerged in the past week?" she asked.

"A couple." I saw where she was headed, but only lifted my shoulders. "If I don't use them, who cares?"

"More than the week before, yes?" She didn't wait for an answer. "Today I have seen you bare them twice. It will grow harder to control the rage with each passing day now. In the end you will be in a permanent state of fury. It will not matter if someone touches Kol. That they stand close to him, or gaze upon him, will be enough to trigger your territorial instincts. You will attack, and you will kill."

"I'm getting a serious migraine here." I pressed the heel of my hand against my forehead, then dropped it and met her gaze. "Let's say you're right about this rage thing. He doesn't want me, Sparky. I can't force this on him."

She didn't blink. "You can if you declare your Choice openly before the HouseClan."

"The Jorenian version of a shotgun wedding. Who knew?" I shook my head. "I can't do it. Even if it means someone has to kill me to keep me from going nuts."

"Why? Why sacrifice yourself and Kol? Do you know what happens to the warrior-bonded of those who embrace the stars?"

"You bond for life, I know. So he'll be lonely." Actually, I could enjoy the thought of that right now.

"No. Without Choice, Kol faces the emptiness of eternity. He will have no alternative." Sparky rubbed her temple. "Your death will take him with you."

Bek sent me back to the infirmary to have my sutures repaired, then dismissed me from training for the remainder of the day. I paced our quarters for about an hour, too tired to sleep, too angry to sit still. The thing inside me that wanted blood wasn't going to let me rest until I spilled some.

I found myself heading for Fayne's rooms and decided to follow my instincts. Bypassing the panel controls was next to impossible, so I waited for one of her roomies to show up.

The oversize red bug appeared, limping, and released the door. I waited a beat, followed him in, and hit him over the back of the head.

He went down with a whomp, and stayed there.

"Pizza delivery." I looked around corners. "Anyone order a double cheese with extra pepperoni?"

No one else was around. While I waited for Fayne, I amused myself by searching her rooms. She had a lot of interesting garments and food stashed, as well as a few weapons. I wondered how she'd smuggled them in, then decided she had to have connections within the Tåna. Finding out who that was, however, proved unsuccessful.

As I crossed the room, the bug buzzed and lifted his head a few inches. I brained him a second time before I heard voices and slipped behind a garment storage unit. From where I was, I could see the whole room.

Fayne entered, secured the panel, and turned to Kol.

"We can talk here," she said, her voice low and amused.

"What about him?" Kol nodded at the bug.

"Cheev, you lazy roach. Get up." Fayne kicked him, but he didn't move. "Hmmm. I hadn't thought him that badly injured from his bout."

"We should signal medical." Kol went to the room console, but Fayne caught his arm.

"Wait." She pressed herself against him, wriggling her torso for a better fit. My claws punched out of the ends of my fingers, and I jammed one heel into the top of my other foot, trying to suppress the beast. "You don't have to play this game with me, Jorenian. Your pet Terran isn't here now."

His pet Terran was going to personally inspect some skink intestines in a few seconds.

"I thank you for your compliment, lady." Kol took her hands from his neck and pushed them down at her sides. "Yet I must decline again."

"Are you still concerned with those cultural taboos? No one has to know, I promise you." Fayne unfastened and shrugged off her shirt. "I was made to fight and pleasure, Kol. I am weary of fighting now." When she would have thrown herself on him, he stepped aside. "You cannot mean to deny me again."

Slowly the red mist dimming my vision cleared. *He declined. Again.*

"You offered to form an alliance with the white neutrals. That is why I came to you last night, and why I am here now.

You have spoken at length about many things, particularly yourself, but nothing of truce." Kol folded his arms. "I am waiting."

The Skogaq's eyes narrowed. "And what if I told you I would not form an alliance with the white neutrals unless I have you for my pleasure?"

"Then I must offer my regrets, and leave." He went to the door panel.

Fayne intercepted him. "Don't be a fool, Jorenian. Deny me, and you will all die under our blades."

"Is this how you lure males to your bed? Offering one thing, promising another?" Kol shook his head. "I came in fairness to negotiate, Skogaq, but I am done with you." He left.

I silently applauded. *Take that, you little bitch.*

Fayne stood staring at the closed door panel, her dead-white face mottled with an odd, grayish tinge. I nearly knocked over the unit when she threw back her head and screamed without warning, then slammed her fist into the panel.

Although I wanted her dead, I knew just how she felt.

She went over, kicked Cheev a few times, then stalked out of the room. I waited a few minutes before I followed, too smug to be a slave to my inner beast—for the moment, anyway.

Kol slipped into our quarters later that night, and dropped onto his mat as if exhausted. I knew because I couldn't sleep until I saw him walk through the door.

Then when I saw him, I didn't want to sleep.

As soon as I heard his breathing deepen to a regular rhythm, I rose from my mat and walked without sound to where he lay sleeping. In the dim light from the console, his face looked different—younger, less austere.

I knelt beside his mat and watched him breathe, and wondered why that seemed more important than sleeping.

"I did not touch her," he said, his voice barely a whisper, his eyes still closed.

Somehow I'd known he'd be awake. "I know." I simply wasn't going to tell him *how* I knew.

He rolled to his side, resting his head against his arm as he looked up at me. "This becomes more difficult by the day."

He reached out, then drew his hand back. "Being unable to touch you, it torments me."

I glanced over at Danea. "Sparky seems to think we'll end up killing each other if we don't claim this bond or whatever it is." I ran a fingertip along the cut on my cheek. "I think she's on to something."

"Who is Rijor?"

I blinked. "How do you know that name?"

"You say it sometimes, when you are sleeping."

"He was a crossbreed I grew up with on Terra. He watched out for me and my mother when we were underground. He talked me into trying out for the StarDrivers, too." I looked down at my hands. "He was murdered on Terra a few years ago. I loved— I honored him a lot."

Kol frowned. "Was he your Choice?"

"No. It wasn't like that."

"But you had relations with him, did you not?"

I could have gotten up then and gone back to my mat. "Yes. We had sex. There were a couple of others, too, before him." I sensed his disapproval. "That's how it works on my homeworld, Kol. You don't have to get married, or have kids, or spend a lifetime together. It doesn't have be eternity. You can just enjoy the moment."

"I have never wished I were Terran, until now." He put his fingers against the wound on my cheek. "Your pardon for this, lady."

I wanted to sink into him, drown myself in his scent, suffocate myself with his flesh. "None is required, warrior." In that moment, being so close to him became more than I could stand, and I pushed myself to my feet. "Good night."

"Sajora." He curled a hand around my ankle. "Had you and Kalea stayed on Joren, I would have found you. I would have Chosen you."

"You keep saying stuff like that to me, and then shoving me away." I moved my leg, and his hand fell away. "If you don't want me, damn it, don't talk about what might have been."

I didn't sleep for the rest of the night, but found some cold consolation in the fact that Kol didn't, either.

Several days passed, during which Kol again spent most of his time away from training and us. I ignored the desire to hunt him down and act in less than a sisterly manner, and instead applied what I'd learned from the experiment with Os and the Tingalean. With a little hunting, I found willing tutors for the rest of the clan, though it took some effort to convince the others the idea would work.

"I know you don't like hurting anyone," I told Nalek as I took him to meet Alvred, a black-skinned e'Pafostol who was half again as big, and twice his weight. "Alv says it's inherent to your species' nature."

When they met, Alv bowed formally and studied Nalek. "You are small."

Nal grimaced. "I often wish I were smaller."

The e'Pafostol turned his back on him to address me. "This runt fears himself more than an opponent. Why did you deceive me?"

I held up my hands. "Hey, I just thought I'd give it a shot."

"Then you should not think." Alv pulled his blades and nodded toward the quad. "You should dance."

Nal tried to step between us. "There is no need for a challenge."

An e'Pafostol hand landed in the center of my ClanBrother's chest and shoved him back three feet. "Stay out of this, runt."

"I'm not fighting you over a simple misunderstanding," I told Alv.

"You believed this coward worthy of my tutoring, did you not? And yet look at how he cringes at the thought of spilling blood." Alv made a rude gesture over his huge shoulder. "Come; we will dance, you and I."

"No," Nalek said, tugging at my arm. "You will not do this, Jory. Let us leave."

Alv casually clipped my ClanBrother across the jaw with his fist and sent Nal sprawling on the floor. "I told you, runt, don't interfere."

"Better go, Nal." I started for the quad.

Nalek picked himself up and went after Alv again. "Leave her alone."

The e'Pafostol buried his fist in Nal's stomach. "Are you deaf as well as puling-hearted?"

The big, dark green body folded in half; then something wonderful happened. With a strange animal growl, Nal launched himself into Alv headfirst.

I smiled with pride as I watched them roll on the floor, grappling and punching each other. "Very nice. Oh, good one, Nal. Keep it up. That's the way. I'll see you guys later." I started for the trainee corridor.

"Jory?" Nal thrust Alv off him and staggered to his feet. "Are you leaving me with this madman?"

"I don't have to watch him train you the whole time, do I?" Then I winced as Alv tackled him from behind. "Good luck, big guy."

There were no crystalline life-forms at the Tåna, so it was harder to figure out who would be *like to like* for Renor. Finally I approached a female trainee named Phehaa, who agreed to meet with him and show him some of her moves.

"She does not appear very dangerous," Ren commented as he entered the session room and saw the humanoid female waiting.

Phehaa nodded to both of us, then took off her tunic and an insulating garment similar to the one Ren wore when training. As soon as she bared her skin, thousands of sharp, six-inch hollow spines sprang up.

"Dangerous enough for you now?" I asked him.

"Ah, that feels better." She sighed and ran her prickly hands over her spiky derma. "I am Emsalmalin," she told Ren. "Physical contact of any duration with most other species is impossible for my kind. Saj tells me you suffer the same inconvenience."

Ren nodded.

"On my world, we are taught *ghnilaatp*, the forms of self-control in dominance and submission. Like so." She walked up to me and made as if to throw her arms around my neck. She stopped so suddenly that her spines hovered a millimeter above the surface of my skin. "As children we learn to focus our energy on the body disciplines, to prevent harm, and to inflict it." She eyed Ren. "There is more that troubles you than your outer being."

When he glanced at me, I shook my head slightly. I hadn't mentioned a word about his other talent.

"It is well, my friend, that you train with me." Phehaa nodded to me. "As I told your comrade, I can give you the discipline and focus you need for the inner and outer being. I would be interested to learn how you remain so . . . untouchable . . . without *ghnilaatp*."

Renor didn't look happy when I left him with Phehaa, but then, he never looked happy any other time.

Sparky proved the most resistant.

"I do not need remedial training." She walked right past the Imabjaic I'd asked to spar with her. "I will overcome my difficulties on my own."

I knew just how to get to her. "I see. So you can dish it out, but you can't take it."

Yellow hair began to rise. "What say you?"

"Hwitloc has a biofield, too. Show her, Hwit."

The Imabjaic strolled over, pulled off a glove, and used a tendril sticking out of the back of his fin to tap Danea on the arm. The two fields clashing sent a small burst of sparks into the air.

Sparky herself jumped about a foot off the floor, then rubbed her arm. "You shocked me!"

"I haven't your natural coronaura, but yes, I can produce a fairly intense charge." Hwit grinned, showing piranhalike teeth as he tapped her again, sending her stumbling back. "In the reefs, it is a matter of survival. I would enjoy sparring with you—your kind are prized trophies on my world."

"I am not a trophy," Sparky said, and stalked into the session room. Hwit nodded to me and followed after her.

I felt like I'd accomplished a great deal when I headed back to our quarters to get some sleep. The sound of blades clashing from another room made me glance inside.

Kol was sparring with a blade dancer. Not just any dancer, either—he was fighting with Uel, the Blade Master.

"HARNESS AMBITION AND IT WILL FUEL THE LENGTH OF YOUR JOURNEY."

—TAREK VARENA, CLANJOREN

"I need a word with you," I said to Bek when we had finished the bladework session the next day.

"I cannot provide you with more private training rooms," he warned me as he dismissed the other trainees.

"How about the one Uel and Kol are using?" I watched the surprise flicker in his gaze, and nodded. "I saw them sparring last night. Since when does the Blade Master provide personal training?"

"Since Kol exceeded the skill levels of every trainer he has," Bek said.

That made sense, but I was still suspicious. "Why not advance him to the order, then?"

"Kol, like the rest of you, must still undergo the Tåna-Shen." Before I could say anything, he held up a paw. "It was by his request that he has been held back all these weeks, so that when the time came, you could fight together."

That definitely sounded like Kol. "I see. So he's been training exclusively with Uel all this time?"

"The Blade Master does not inform me of his daily schedule, but yes, I believe they have been sparring together frequently of late."

"You might have asked me, Saj." The Blade Master stepped out from one of his tricky hidden recesses and gestured toward the arena. "For now, however, you will assemble with the others around the quad."

"I didn't hear a summons," I said. "And I'd like to ask you a couple of questions."

"Ask later." He touched something on the wall panel, and a drone voice called third level to assemble. "Go now."

Bek and I followed Uel out to the arena, where Dursano was waiting inside the quad. The Blade Master's appearance silenced the assembled third-level trainees at once, and everyone watched as Uel entered the quad and Dursano placed a small projection unit at his feet.

"Before now, nothing of the outside worlds you came from has been allowed to interfere with your training. However, certain extraordinary events have occurred which must be discussed."

Uh-oh. This doesn't sound good.

"We have been notified by the Hsktskt Faction and the Allied League of Worlds that this region of space has come under immediate occupation by forces from both sides." Uel nodded to Dursano, who activated the projection unit. A dimensional image of Reytalon and the surrounding system appeared, wedged between two immense fleets of military vessels.

There were more than three thousand ships out there, and we were sitting right in the middle of them.

"Mother of All Houses," Nal murmured beside me.

"League and Hsktskt forces have effectively blockaded Reytalon from all trade and supply routes, placing the Tåna fully at the mercy of their warring armies," Uel said. "In addition to this, leaders from both sides have transmitted specific demands to me."

Dursano projected a split image, one of a particularly nasty-looking Hsktskt OverLord, and another of a League general.

"OverLord CulVar of the Hsktskt demands all Tåna trainees be transported at once to his vessels, where they will join faction forces and combat the League forces." Uel paused for a moment. "General Hughes has ordered all trainees be inducted into the Allied League forces to do the same to the Hsktskt."

"Back to the rock and a hard place," I muttered.

The other trainees made some noise, too, until Dursano held up his hand.

"I have no intentions of surrendering any of you to either force," Uel said. "That is not the way of the order."

"You haven't got a choice, Blade Master," I said, loud enough to be heard. "If you don't, they'll invade."

"I have offered a compromise to both coalitions, to which they have agreed. All remaining second-level trainees will be advanced, and along with you will be permitted to complete the training and join the order before you are appropriated." Over the sound of shocked voices, Uel added, "I have no intention of keeping that compromise, I promise you. It is merely a tactic. I have begun making separate arrangements to provide safe passage for everyone from Reytalon. I know all of you will recognize this next image."

Dursano projected a replica of the spacial anomaly that had sucked us in during the battle with the Garnotan ship, only brightly colored gaseous clouds surrounded this one.

"This is the Schaller Rift, through which all of you were brought to Reytalon. The location and nature of the phenomenon is unknown to anyone outside the order; while ships may enter and exit our solar system through conventional routes, the rift lies camouflaged within the remains of the Tnekar Nebula. The rift itself allows undetected passage in and out of our system. It is the reason the order chose to build the Tåna on this world.

"The rift is the escape route we will use. Members of the order will soon arrive with ships to take all of us out of the war zone. Until they reach Reytalon"—Uel gestured toward the quad floor—"we will continue as if we mean to keep the compromise with the Hsktskt and League. Those of you who wish to stay behind and join either force will not be compelled to leave with the order."

"How are we to be certain you will keep your word?" I heard Renor call out.

"I can swear before any deity you choose, but nothing is certain beyond death, Ren." Uel seemed amused. "However, if you wish to verify my actions, I will provide access to the Tåna database to a representative for the trainees."

"Silvers do not need verification." Fayne moved to stand beside the quad. "The word of the Master is enough for us."

A Cordobel emerged from the crowd. "Browns trust in the word of the Master."

Everyone looked at us.

I was tempted to say the whites wanted everything etched on crystal, but Kol stepped forward. "White neutral accepts the Blade Master's word."

"Then it is decided. Training will continue as scheduled, and all trainees are advanced to purple. The Tåna-Shen will be held in two weeks. Prepare yourselves for the final challenge."

Being in immediate danger changed everything at the Tåna, and not for the better. Uel's offer to get everyone off Reytalon before the Hsktskt or League could invade should have acted as a buffer between the silvers and the browns, but in a strange way it only made things worse. Being denied the opportunity to immediately join the occupying forces infuriated some, while others crowed over which side would prevail in the war. Soon the quad never remained empty for long. Silvers and browns constantly challenged each other to bouts, and sometimes fought to the death.

In the center of the growing unrest, the white neutrals were fair game for either side.

Then there were all the remaining second-level trainees who were advanced to third so they could participate in the Tåna-Shen. They were like walking target forms, and Fayne decided to vent her spleen on them by directing her silvers to drag the ones who would not join them into the quad at any opportunity. Four died during one session alone, until Kol brought most of them under the protection of the whites.

"If you wish to survive to escape, you must pair off with the more experienced," he said during an impromptu gathering in the arena. "Each of you will be expected to do the same as any white—do not challenge, and do not accept challenges. Protect your partner at all times. And remain neutral in any discussion on the war."

Despite absorbing the second-level trainees, white neutral remained outnumbered by both silver and brown, three to one. Pairing off didn't always help, either, as evidenced by a challenge accepted by two whites, who were subsequently slaughtered in the quad by Cirilo and Cheev.

"At this rate, Uel won't need escape vessels," I muttered to Ren as I watched the bodies being dragged from the quad. "We'll kill each other long before they or the bad guys get here."

"What would you suggest I do, Sajora?"

I turned to look down at the shrouded face of the Blade Master. "Deactivate the implants. Better yet, have them removed."

"The implants control the transmutation point of the tåns. If they are removed, your blades will remain solid."

How convenient. "Then get rid of the poison trigger."

"That would require removal of the implant."

"You've got an answer for everything, don't you, Blade Master?" I nodded toward the two dead trainees. "Except they can't hear you anymore."

"Do you wish to be a dancer, or a philanthropist? For I assure you, Sajora, you cannot be both." Uel strode off.

"First person I challenge when I get the black," I said to Ren, "is him."

"Considering the amount of time the Blade Master spends watching you, I would advise against such a course of action." A halo of light glittered around his face as he turned to scan the arena. "I have not seen Danea. Does she practice with the Imabjaic this morning?"

"No." I looked around, too. "They were only sparring after training, I thought." A minor commotion was in progress on the other side of the arena—more silvers and browns posturing for challenges—and then I saw a familiar glow, and someone fly backward. "Shit."

Ren and I went over to find Sparky and Hwitloc back-to-back in the middle of a widening circle, their garments torn, their blades unsheathed. Someone had cut Danea's arm, and Hwit's face was bleeding from two shallow wounds.

Someone being Cirilo, judging by the blood on his hands and blades.

"You're having a party and didn't invite us?" I said as Ren and the whites fanned out behind the silvers. "I'm so hurt."

Cirilo's recessed eyes swiveled toward me. He was grinning. "Does cow go with fish?"

"I'll ask your mother." I didn't take my eyes off him. "Sparky, want to tell me what's going on?"

"The pinheaded one cut Hwit as we entered the level. The rest came after." Her glow increased. "Most of them have insulated themselves somehow."

I made a *tsk*ing sound. "Pinhead, Pinhead. Now I'm disappointed." Out of the corner of my eye, I saw Nalek and the rest of the clan heading our way. About time. "Say good-bye to the third eye."

"Cirilo." Fayne emerged from the pack of silvers. She didn't look happy, either. "What are you doing?"

"Her." He pointed a blade at me. "I want her."

Fayne gave me an odd look. Not the usual, I'd-like-to-see-you-dead glare. No, this was different. Creepier. "I have prior claim."

"Sajora." Kol hovered just behind me. "Walk away."

"You get the feeling Mommy and Daddy don't want us to fight?" I smiled at Cirilo. "I say we take it to the quad."

"He's mine first." Sparky stepped in front of me. "I challenge you."

Pinhead nodded, turned, and headed for the quad. "You can be second, Terran."

"How much control has she gained?" I asked the Imabjaic as we took position below Danea's corner of the quad.

"Not enough." Hwit's eyes rolled as he glanced around us. "She still projects without thinking when she spars, which tires her, and that silver's thermals will protect him throughout the match. I fear for her, Saj."

Across the arena, I saw Fayne watching me. "Keep your fins crossed."

Danea left off her obek-la, and her hair was radiating in all directions as she stepped to the center of the ring. Cirilo strolled with the confidence of someone assured of the win.

"Something is not right, Jory." Renor came to stand on my left. "The Fawgithin prevails through deceit."

"We won't let him." I'd jump in the quad myself, if I had to.

The hoverdrone descended and initiated the bout. Cirilo attacked with a lateral right, which Sparky countered easily. What she didn't anticipate was the follow-up body slam, which knocked her back two feet, nearly into the ryata. A bright yellow light crackled in the air.

I'd seen Sparky glow, but never flash. "What the hell was that?"

"He's wearing a neutralizer," Hwit said. At my blank look, he explained, "It's a device used to drain bioenergetic fields. The Hsktskt use them when they attack aquatic colonies."

"Figures." I looked at Ren. "Want to help me break up a fight?"

"You will not interfere." Dursano and a couple of inductors surrounded us. "The neutralizer is allowed."

"So he can suck the energy out of her, then kill her? I don't think so." I glanced down as the inductors drew their blades. "She's my family; I'm not going to watch her die."

"You will watch, or you will die."

"Jory." Ren put a cold hand on my sleeve. "We all have our talents. Trust that they will see Danea through this match."

I stared at him, then recalled Uzlac's mysterious end. There was another flash from the quad. "I hope they hurry up."

Ren went very still, and concentrated on the fight. I stood beside him, itching to pull my blades and jump in. Then I felt a strange, chilly sensation in the air around me, and a sense of something collecting and growing.

Hwit raised a fin as if to feel the air. "What is that?"

"Someone adjusted the temperature controls again." I gave Dursano, then Ren, a deliberate look. "Right?"

He nodded slowly.

Whatever Ren was doing, it was making his body temperature drop, too—the hand on my sleeve felt like a block of ice. In the quad, Danea was circling and panting as she avoided Cirilo's casual jabs. He wasn't even making an attempt to spar with her, only chased her, trying to make contact so he could drain her further.

Danea's hair spiked as she turned her head for an instant and looked at me and Ren. Then she stopped and spread out her arms, leaving herself completely defenseless. Cirilo laughed, lowered his blades, and jumped at her.

What happened then nearly blinded all of us.

The entire arena fell silent as the yellow flash became twin streams of unbearably brilliant golden spheres that poured out of Danea's hands, joined together, and shot out to slam into Cirilo's unprotected face. He stopped in midlunge, almost suspended in the air as the energy poured over him, then slowly rose until his feet left the quad. He kept going up, higher and higher, buoyed by the spheres, until he was twenty feet above the quad.

Ren's grip on my arm relaxed, and the spheres burst at the same time. The yellow light vanished, causing Cirilo to drop like a stone and hit the quad face-first. He twitched once, groaned, then didn't move again.

In the quad, Danea staggered to the ryata and held on, panting. Stunned voices rose in confusion around the arena. The hoverdrone descended to inquire if Cirilo wished to continue.

Pinhead didn't answer, and when the drone rolled him over, he stared up with wide, lifeless eyes.

Sparky needed a couple of hours in the immersion tanks to recover from the stress of the bout, and Renor and I stayed in the infirmary to keep her company and keep other, curious trainees away from her.

"She sure knows how to fry someone in the quad," I said to Ren as I watched her swim. "Mind telling exactly how you helped her do that?"

"I did not help her. Danea was merely able to successfully encapsulate her field," he said.

"But Cirilo was wearing that gadget; why didn't it work?"

"I believe they will find it malfunctioned."

I rubbed my temple. "I thought your push only worked on living things."

"The energy it was draining came from a living being. I used the energy as a conduit." He smiled at me. "Did I do well, ClanSister?"

"Yeah." I chuckled. "You did fantastic."

Fayne made a big deal over Cirilo's death, of course, which infuriated the silvers and compelled the balance of the trainee population to take sides. Many chose to become white neutrals, but we were still outnumbered on both sides.

The silvers stepped up their campaign of terror, brutalizing anyone foolish enough to get in their way out of the quad, and butchering anyone they could lure into it. The daily mock battles became more vicious than ever.

And still Uel did nothing to stop it.

"It's like he wants us to kill each other off before the rescue ships get here," I said to Os on our way back from training one evening. "I thought the whole idea here was to *save* everyone's ass."

Inside our quarters, Os shrugged off his tunic and stretched his other arms. "Perhaps Kol will speak with him." He saw Galena limping out of the cleanser wearing only a towel, and took a step toward her before catching himself. "Or I will."

I went over to look at Birdie's leg, which sported a hideous bruise and a long graze. As far as I knew, she hadn't been in any of the day's melees. "What happened to you?"

"I fell." She looked over my shoulder at Os. "It was an accident, my ClanBrother."

Os got an ugly look on his face; then he pivoted and strode out.

I crouched down to check her muscles. "It's going to hurt for a couple of days, but you'll be okay. Stop flattening your wings. What kind of accident?"

"I . . . reacted improperly to a diving maneuver Strix demonstrated for me." She flexed her wings. "It is difficult for me to think like a predator, even with regular practice."

I made a mental note to thump Strix the next time I saw him. "So he did what? Clocked you in midair?"

She grimaced. "Something like that."

We both looked up as Kol and Nalek entered. They were speaking in low voices, but abruptly fell silent when they saw us.

"Don't mind me and Birdie," I said.

Kol looked at Galena's leg, then around our room. "Where is Osrea?"

I helped Galena hobble over and stretch out on her mat. "I don't know; I'm not his mother."

Nalek rubbed his bald head. "Renor and Danea are using a practice room. He will be alone, wherever he is."

Kol turned and walked out.

"You stay here with Birdie, big guy." I went after Kol, and caught up with him just outside third level. "Didn't you give him the stay-in-pairs speech?"

He scanned the arena. "Osrea is impulsive. As you are."

"And you're not?" I got a strange feeling and looked up. "Oh, hell."

Strix and Tej were flying overhead, high up in the dome, carrying a writhing Os between them. Their talons prevented him from using the blades in two of his hands, but he was doing a great job of plucking them with his other pair.

"Strix!" I waved. "Would you bring him down here? In one piece, please?"

The two avian beings slowly descended and dropped Os about five feet from the floor. He landed, rolled, and leaped to his feet. "I'm not done with you!" he shouted at the avatars.

"You are now." I clamped a hand on his shoulder. "What do you think you're doing?"

He whirled on me, baring fangs I'd never seen before. "They *injured* Galena."

"As part of training, yes, they did. We've all gotten hurt plenty of times since we came here. And just for the record"—I tapped a finger under his plated chin—"if you bite me, Snake Boy, I get to bite back."

"Osrea." Kol made a gesture, and the two of them walked away from me. More conversation in low tones. That was really starting to bug me by the time they returned.

"Why is it no one wants to talk around me anymore?" I asked with a sweet smile.

Osrea's tongue flickered. "Because you pick fights."

"That's rich, coming from you. Come on." I sighed and headed for the corridor. "Let's go get some sleep."

Snake Boy went into our quarters ahead of us, but Kol stopped me by putting an arm across the doorway.

I glanced down at it. "I'm not in the mood for another debate on Jorenian mating practices, if that's what you have in mind."

"I have discovered some information about Kieran." He nodded down the corridor. "Walk with me for a while."

I walked beside him, and waited. And walked. And waited. "Kol, much as I appreciate the complete guided tour of the trainees' quarters, I'd like to get a couple of hours' sleep tonight."

He stopped at a door panel and opened it. "In here."

It was a room I'd never noticed anyone going in before, and saw why when I stepped inside. Databases and consoles lined the walls, all active and waiting for someone to tap into them.

"How did you get the code?"

"I observed Uel entering, and memorized his input." He went over to one of the consoles and tapped in an inquiry. A long list of data scrolled onto the display. "These are attacks on various ships, colonies, and space stations attributed to Kieran."

I examined the list. "There are thousands of them. But that can't be right. The League database barely covers a tenth of these reports. Where are they originating from?"

"Here, at the Tåna." Kol down scrolled until he revealed a line of numeric code. "They have been keeping detailed records on Kieran's activities—and those of every other blade dancer in the order—for a very long time."

I rubbed my brow. "For what purpose? Blackmail? An honors list? Payroll?"

"I do not know." He indicated the most recent entry. "According to this, Kieran returned to Reytalon, but there is no record of his departure."

That hit me like a slap. "So he could still be here. Or he died here."

"Both possibilities occurred to me as well."

"He can't be here. The only Terrans are you and me. Bek said so." I groped for a chair and sat down. "We are the only ones, aren't we?"

"I asked the same of Uel. He indicated we are."

I read over the entries. "Why would the Tåna track him like this? He wasn't a blade dancer; he was a raider. Why would they care?"

Kol went behind me, then rested his hands on my shoulders. "Sajora, Kieran never left the order. I do not believe anyone ever does."

"So he's still on active duty?" I laughed, and leaned my head back to look up at him. "Sorry, I was just thinking, could this get worse?"

"Bek admitted to training him. We know he graduated and joined the order."

"Yeah, but thirty years ago. And he hasn't exactly stuck by the old order motto of 'Kill and don't get caught.' He's been openly raiding anything that moves." I wished I could have told him the rest—that Kieran was my biological father—but I didn't want him to know that it had been my father who had sold his mother to slavers.

Kol's fingers began stroking the tight muscles on either side of my neck. "Sajora, there is something more to this than a collection of data. Look at the date when Kieran returned."

I looked. It was only a few weeks before I'd been deported from Terra.

"It cannot be a coincidence that we are here as well."

I was going to ignore the wonderful magic his fingers were working on my neck. Somehow. And I was going to change the subject before I ended up blurting out the truth. "I don't think Kieran has anything to do with that."

He turned me around in my chair, and knelt before me. "I am only saying that it is very suspicious." He moved his hand, brushing some curls away from my throat. "Your hair grows quickly."

"So does yours." I chased his straight black strands with my fingertips, enjoying the way they felt against my skin—a lot like heavy dimsilk. "What do we do now?"

"We must discover who among the Tåna staff is involved with Kieran, or knows of him. Bek would be the obvious choice." His white-within-white eyes looked all over my face. "You are tired. We should return."

"No." I let my hand rest against his chest, over the implant that could kill him in an instant. "I want to stay here a little longer."

"We agreed." He swallowed as I brushed my thumb across his jawline, then tried again. "We said we would not do this."

"Do what?" I murmured, watching his lips move, unable to make sense of the words.

"This." He cradled my face with his palms. "This." He kissed me, a whisper of mouth against mouth. "This." He rested his brow against my cheek, and breathed in.

I felt the inhuman power rising inside me again, but it was much different this time. It had become huge, and hungrier than ever before, but it brought with it heat and desire, and the need for more than physical satisfaction.

"Do you feel it?" he whispered.

Yeah, I felt it, all right. I didn't want to have sex with Kol. I wanted Kol. All of him. Every breath he took, every day he lived, I wanted it for me, and I wanted to give him mine. My life for his.

Finally I understood what Kol felt, what his people felt when they Chose. I wasn't so Terran after all.

It should have been simple. An exchange of words, a promise of bonding our lives together. I could almost hear the words in my head—*I will be yours, and you will be mine, and we will walk the path together into eternity. Sharing everything between us.*

Yet as much as I wanted to say those words, and hear them from him, I knew in whatever part of me that was still Terran that it could never be that simple. Kol was a warrior, an honorable man who believed in what his people had taught him. Despite the way they'd treated him. In spite of everything I'd told him.

And what was I, compared to that kind of nobility? Far less than he deserved. Maybe it would have worked if I'd grown up the way he had, on my mother's homeworld, instead of living like an animal on Terra. No matter how Jorenian I felt in that moment, I could never be the woman he wanted.

And how would he introduce me to his kin? *Hi, this is my bondmate, Jory. She gets her green eyes from her father—you know, the guy who put my mother up for auction.*

"Kol."

He lifted his head, blinded by the same needs I was feeling, lost in the same haze. The only thing that kept me sane was knowing how helpless he was before me, under my hands, completely riveted by this thing between us.

He'd been the strong one in this, until now. And because I loved him, I would have to take it from here.

I kissed him one more time, because I couldn't help myself. Then I pushed his hands away and got to my feet. "We're not doing this." My voice shook. "You've got to help me out here, okay?"

He caught me from behind, his hands clamping around my waist. "What say you if we went on, as Terrans do? As you have done with the others in your past?"

No, he wasn't going to be very helpful, damn him. "You aren't like the others. I could walk away from them when it was over."

He slid his hands around to splay them over my belly, and nuzzled the back of my neck. "Perhaps you would not wish to leave me."

He wasn't going to leave me alone unless I did something dire. Like wake him up. "Maybe. Or maybe we'd end up trapped together forever, like Qelta and Nla."

Kol lifted his head, and took his hands away. "What say you?"

I turned. "You've got to know how unhappy they are together." He didn't blink. "Why do you think Nla spends all his time out in the fields? It wasn't all you, Kol. Why did he insist that the three of you never live with HouseClan Varena?"

"The Varena adhere closely to HouseClan law and protocol—far more than any other House. Nla may have been trying to protect us."

"You mean hide you. The only thing I don't know is who he hates more—you or your ClanMother."

"They Chose. They share the bond."

"Oh, yeah? Then how come they can't stand to touch each other?"

His eyes became slits. "They do not indulge in public displays of affection."

"They weren't in public the day we left." Hurting him backlashed on me, and my claws emerged. I'd never realized they could come out in pain as well as rage. "I don't want to end up like them, Kol. Hating each other, tied to each other, no escape, no relief."

He got all formal and Jorenian on me. "As you wish, lady." Then he left me standing there, staring at the list of my father's crimes.

CHAPTER EIGHTEEN

"DIRECTION TAKES MANY FORMS."

—TAREK VARENA, CLANJOREN

We were in bladework, running through the last hundred or so moves to complete our *shahada* repertoire, when another trainer appeared and interrupted the class.

"Emergency signals have arrived for the following trainees." He read off a list of everyone in our group but me. "You will accompany me to the communications area."

"What's going on?" I asked Kol as he passed me. "Is Joren under attack?"

"I will find out."

I was being deliberately separated from the clan for a reason. But what? "Let me know."

Bek ordered the rest of us to get back to practice, but hardly an hour passed before third level was summoned out to the central arena to assemble around the quad. Then something very weird happened—someone bumped into me as I was making a final inside lateral block, which allowed my sparring partner to give me a small nick on the back of my hand.

"Saj, you should go to the infirmary now," the Chakaran told me, holding me back.

"It's just a scratch." Bek sounded worried, which got my attention. "What is it? What's happened?"

He didn't answer, and a moment later the Blade Master materialized beside us. "Sajora will enter the quad."

Bek glowered at Uel. "She cannot. She is injured."

So that's why he wants me to go to the infirmary—so I don't fight. "I said I'm fine." An eerie sense of the inevitable crept over me. "Who else is in the quad, Blade Master?"

"An opponent eager to test her blades against yours." Uel pointed to the door panel. "Go and see for yourself."

Inside the quad was Fayne, doing more of her razzle-dazzle, spinning her blades and moving through the most complex of *shahada* patterns to impress the gathering crowd.

The beast inside me—the one I'd been suppressing for weeks—rolled over, sat up, and took notice. Internal chains stretched and groaned.

The oohs and aahs stopped as soon as I mounted the platform, and basically everyone on third level turned to stare at me.

While I stared at the broken body being dragged from the quad floor. The avatar's face had been mutilated, but I could tell it was her from the tunic and the broken wings.

"Come, Terran." The little white rat came over to lean against the ryata, and tossed a couple of small objects at me, which I caught out of reflex. "Let us dance."

I looked down at the eyes, smeared against my palm like two broken bird eggs.

Galena's eyes.

Everything inside me snapped, all at once, and my claws emerged from the ends of my fingers so fast and hard they tore the skin. Other trainees moved out of my way as I pushed my hand inside my tunic and pressed the ruins of my little sister's eyes against my heart. A sound came out of me then, a howl of despair and outrage.

For this, she dies, the beast inside me whispered. *Slowly.*

I ducked under the ryata, my blades ready, my gaze locked on Fayne's sneering face. I could have called her names, taunted her, but my throat had been seized up by the beast's claws and I couldn't utter a word.

Blood. That was all I wanted now. To see her blood staining my blade.

A drone unit descended and I used a fist to knock it out of the arena. Fayne smirked as she countered my moves, mirroring me as I took position. "You do not wish to make it official?"

The hilts of her blades slapped into her palms as she stopped spinning them. "I see you do not. Then let us begin."

I stood my ground, letting her attack, watching her limbs bunch and flex as she lunged across the quad. Absently I thought of Terran cheetahs, and how they nearly dislocated their hips when they ran. A second before she touched me, I whipped up my blades and slashed at her face and chest.

Everyone in the arena began to shout.

Something sliced across my abdomen, and the center joint of her arm smashed up under my chin. I pivoted and followed through with a knee to her groin, but she rolled away before it had any effect. A soft snowfall of white hair rained down on the quad between us, a good-sized chunk I'd slashed off her head.

She touched her skull with the back of one hand as she circled to my left. "You cut my hair, you clod."

"Jory!" I heard someone shout over the clamoring voices around the quad.

"Here is your beloved one," Fayne said, crooning. "He will need comfort later, I think. Perhaps I will allow him back in my bed." She thrust a blade low, trying to get under my guard, then reversed the move and cut up and across my forearm. "Should I give him your eyes as a memento?"

"Concede, Saj," Bek called from my corner. "Live to fight another day."

I didn't care how much blood I spilled on the quad floor. I watched her, countering her moves, letting the beast take over completely. The part of me that was Jory receded to a small corner of my mind, and the wet patch over my heart kept her there.

Birdie didn't deserve to die. Not like this.

The noise around the arena swelled like an infected wound, but I focused on Fayne. It paid off when I spotted a break in her pattern and attacked. I used a combination I hadn't tried before, and the stress on my knee made my leg lock for a moment. The blood under my footgear added to the problem, but I found my center and attacked.

As I moved in to bury my blade in her chest, she spit in my eyes. My vision clouded, and my eyes felt like they were on fire.

The involuntary jerk of my head sealed it, and I slipped backward and fell.

No, God! I flailed my arms, trying to stay upright. The shouts around me became jeers.

Fayne jumped on top of me and plunged her blade into my chest. As the holographite dematerialized, it produced a jolt of such intensity that my entire body went rigid.

"Now"—she pulled the tån back to let it solidify again—"you die for me."

I blocked her arm as the blade came down again, but the tip penetrated my tunic and skin enough to score a second hit. The subsequent jolt allowed her to pin my arms down with one hand, and raise the blade again.

I knew I was going to die then, and turned my failing eyes to see the outline of someone big and blue trying to climb into the quad.

"Kol, no!" I heard Sparky shout. "Osrea, Nalek, quickly!"

Something swooped down and knocked Fayne off me, sending her sprawling back onto the quad. I struggled up to my knees, groping for a handhold.

"What are you doing?" I heard Galena demand.

"Make her concede," Nalek's voice said, and four hard hands jerked me to my feet.

Birdie. I squinted at her, still not sure if I could believe my blurry eyes. If she was alive, then—

"Fucking bitch!" I shrieked, fighting Osrea's grip on me, trying to get an arm free so I could slash Fayne's laughing eyes out of her skull.

"You still have a hit left." The Skogaq tilted her head to one side, and licked my blood off her blade with her little white tongue. "Come and dance with me one more time."

Trainers and hoverdrones were suddenly everywhere, and I saw the Blade Master standing behind Fayne, at her corner. And Nalek hauling a limp, blue body out of the quad.

Kol?

"Her saliva blinds," Os said, and locked one arm around my neck. "Say 'I concede' or I will knock you unconscious," he added in a mutter.

The arena fell silent. So silent I could hear the rasp of my own breath.

He wasn't dragging me out of the quad. But my eyes were burning out of my head, and I couldn't see anything of Fayne but a vague blur. One more hit, and I'd be dead.

I lifted a hand and wiped the blood from my mouth. "I concede."

Os tossed me over his shoulder and hurried out of the quad.

I spent a day in medical, getting my eyes treated for corneal burns and cursing myself for a fool.

"If I see you in here again this month, I'm suspending you from training," the horn-headed doctor told me as he shone a light in each eye and put in more drops. "Wait, don't blink. There's no permanent damage—lucky for you, your clan got you here in time—but you'll need to wear shades for a few days. And stay away from Skogaqs. Now you can blink."

"Where *is* my clan?" I asked as the solution trickled down my cheeks.

"Blade Master sent for them. Some kind of hearing. By the way"—he picked up a specimen dish—"I found Exzer eye material smeared on your breast. What's that about?"

"Long story." I got dressed, put on my shades, and headed for third level.

Dursano stopped me before I entered. "You are relieved from training for the remainder of the day."

He wasn't going to get rid of me that easily. "What's going on with the clan?"

"The Blade Master must decide what to do with them, and Fayne." The inductor held up his arm. "You are not invited."

"That's the story of my life, Inductor."

No matter what I said, Dursano refused to let me enter the level, so I stalked back to our quarters, which were empty. I paced until I got on my own nerves, then snapped my head up as Ren came through the door.

"Well?" I went to him. "Where is everyone else?"

"At the ruling, with Kol, Nalek, and Osrea." He studied my face. "Your eyes, are they damaged?"

"They're fine, no thanks to that little white runt." I tapped my shades. "I just have to wear these for a while. What about the boys?"

"They violated Tåna rule by entering the quad during the bout. Fayne did the same by spitting in your eyes." He made an eloquent gesture. "I am inclined to think that Uel will overlook both violations."

"Why did you leave?"

"We were concerned about you, and I volunteered to check on your status." He smiled a little. "Do you find that amusing?"

"No. I think it's . . . nice." I threw up my hands. "Christ, Ren, this was my fight. What the hell happened?"

"Kol saw Fayne attacking you, and lost control. We were only trying to stop him." He took off my shades and inspected my eyes. "The Skogaq played a cruel game with you today. The next time she could make the game real."

And kill Galena to draw me to the quad. "I know."

"Kol has been in a near-permanent state of rage since the fight. The healer gave him a tranquilizer, but it has had no effect. He cannot think clearly." His cheek glittered. "It has been difficult for both of you."

"So what do I do now, Ren?"

"All the years I spent in isolation, I wondered what it would be like to be a part of some greater thing than myself. You and the others have given that to me."

I frowned. "You're not going to cry on me, are you?"

"Another bodily function my sire's species prevents me from enjoying. Jory." He came and carefully took one of my hands between his. The cold angles of his crystal derma made it feel like I'd plunged my fingers into a basin of broken plas. "I honor the kinship we share, yet if I knew I presented a threat to you and the others, I would return to my isolation at once. Do you understand?"

Sure I did. "What about Kol? Will he calm down?"

"Distance may help. I am not sure—but if you continue to

train in such close proximity to each other . . ." He made an eloquent gesture.

If Birdie died because Fayne had it out for me, or Kol, defending a love we couldn't let happen, I'd never forgive myself. "There's no other way, is there?"

"You have always taken the most demanding path." He released my hand. "That is your gift, you know. Of all of us, you are the most self-sufficient, the most determined. You have never hidden behind a lie, or used one as sanctuary."

How little he knew me. "There's more to this problem than Fayne." The door panel opened behind us as I added, "Somehow, she got Uel to set me up."

The door panel closed. "It was quite the reverse, Sajora. I had you set up Fayne." The Blade Master entered and nodded to Ren. "Excuse us for a moment."

Ren glanced at me, and I waved him out. I could see clearly now, but the dimsilk Uel wore didn't seem to shimmer as much. Maybe it was looking through the special shades the doc had given me. In any case, whatever Uel was, he was humanoid.

"I'm not blind." I pushed my shades up so he could see my eyes. "In any way, shape, or form."

"I gave you your chance to defeat Fayne today—"

"You *gave* me?"

"Perhaps it would be better to show you why I am obliged to hide behind this mask." Slowly he removed his obek-la and revealed his face.

His artificial face.

"You're a reconstruct." I'd seen a few of his kind on Terra, before they'd begun using biografting to conceal their drone alloy skulls. None of them wore the synskin he had anymore. "Full body?"

"All but my brain and a few inches of my spine. My organic form was destroyed during an ambush on Skogaqen. I had myself transferred to a drone chassis." He tugged off a glove and a prosthetic hand to reveal a clawlike grappler. "So you can believe me when I say I have no affection for Fayne or her kind."

Reconstructs had originally been built as cheap slave labor

for League colonies. Harvesting neural tissue after natural death seemed to negate any accusation of slavery, until one of the reconstructs proved they were capable of independent thinking and sentient behavior. They were freed, but plenty of species still had their doubts about reconstructs qualifying as living beings. Uel probably would have a much harder time governing the Tåna if the trainees knew he was ninety percent artificial.

I'd go along with the obvious reasons for him hiding behind a mask, but not the bout. "Then why did you send me out there, knowing what she was going to do?"

"I was hoping she might try to spit in your eyes, as I suspect she's done with all the others she's killed. I trusted your skills would keep you alive long enough to have it verified by a hoverdrone, so I could discredit her in front of the silvers." His synthetic face didn't allow much expression, but I could see a little anger. "Your clan got in the way."

"Kol kept me from being blinded."

Uel removed a small vial of spray from his tunic. "I was carrying the counteragent."

"You said she's done it before. Why didn't you bust her for cheating and murdering them?"

"Skogaq saliva remains detectable for only a minute after it enters the mucous membranes." He tucked the vial in my hand. "Keep it with you. I doubt she will make a second attempt, but her kind are unpredictable."

"Thanks." I looked around. "I need to move out of here, get a room of my own. Can you arrange that?"

"Yes." He pulled his obek-la on. "Do you wish to stay in close proximity to your clan?"

"No." I went to my garment storage unit. "Get me as far away from them as you can."

That same night, Uel moved me to the other side of the trainees' quarters, closer to medical and his own offices.

"Kol has discontinued his training with me," the Blade Master said as I finished putting away my scant belongings. "If you wish, we can spar together before your training session begins tomorrow."

"Am I that bad?" I asked.

"No. You are as fast as Kol is. You lack his stamina and power, but your agility compensates for it. With more training, you will become what you were meant to be." He pulled up a schematic of third level on the room console, and highlighted one of the training rooms. "Meet me here an hour before training commences. And here." He handed me my Omorr blade. "Wear this at all times."

Before he got to the door panel, I asked, "What if I have to kill her?"

He hesitated. "That is why I'm training you, Sajora. You must kill her." When the panel opened, Kol nearly slammed into him, and Uel stepped to one side. "I will see you tomorrow."

The door closed, leaving me alone with the last person I wanted to see. And he was angry, so angry that his skin had flushed a deep blue over his cheekbones and his eyes practically scorched the air between us.

He made a full three-sixty turn before asking, "What is this?"

"This is my new room." I gestured around me. "Like it? Not as big as the old one, of course, but I don't have to share the lavatory or listen to Os snore half the night."

"Why did you leave us?"

"So none of the clan gets made into bait." As if touching a magic talisman, I curled my fingers around the hilt of my Omorr blade. "I don't want to watch you go berserk again, either. This is getting pretty bad, this thing between us. I thought we could use a break."

His expression turned ugly. "If I were Terran, you would honor me."

"Kol, you are part Terran, and I do honor you." I tried to keep the hitch out of my voice. "Now honor me and get the hell out of here."

"I want to see your eyes." He came toward me, still bunched up in knots, and somehow I held my ground. I couldn't help the flinch when he took off my shades, or the shudder when the tips of his fingers skidded down my face. As if he were blind, and it was the only way he could see me.

I pulled his head down to mine and closed my eyes as soon as our brows touched. "If there were any other way, I swear to you, Kol, I would. If there were."

"The others say we are warrior-bonded." He lifted his head and pressed his mouth against my brow. "I believe it may be so. Today I nearly killed three people trying to get to you in the quad. One of them was Nalek."

I swallowed. "He said it only happens on the battlefield."

"We are fighting for our lives every day. As well as our sanity." Self-derision tinged his voice. "If that is not the field of battle, then what is?"

I curled a hand around the back of his neck, trying to control my breathing. "When this warrior thing happens, what do your people do?"

"We fight together as one." He took my hand and threaded his fingers through mine. "We do what we must to survive, always together, so that we may preserve each other and live to bond." He brought my wrist up and pressed it against his throat, so that I could feel the heavy pulse there, echoing the frantic beat of my own. "When there is a lull in the battle, we claim all we can have."

Claim. That was a strange way to put it. Like it was some territory to be defended. "Even with your own sister?"

"The Mother would not allow such to happen." His breath warmed my face as he unfastened my tunic. "You feel that as much as I do."

I would have been happier with a DNA test. "Kol, I know what I want. I know what you want. It doesn't work." It was so hard to drag the words out with his body so close to mine. I hissed in a breath as he pulled off my tunic, then his. Our bare skin touching made me forget my own language. "We agreed not to do this."

"I am not your brother." He bent down and kissed me, his mouth fast and hard and a little cruel. "Say it for me."

"I don't know—" I got kissed again, and felt my trousers pool around my ankles. Then my undergarments. "Boy, that is really not fair."

"Release me, my heart." The agony in his voice matched the expression on his face, the tension in his touch. "Say the words and free me from this torment."

I knew what I should have said. But the memory of Fayne's blade descending on me made me give up. "You're not my brother."

With a growl or a laugh—I couldn't tell which—he lifted me off the floor. That was when I realized we were both naked. "I Claim what is mine."

The room began to whirl, taking me and Kol along with it. Nothing felt familiar. It startled me—I'd had sex before; *he* was the virgin—but none of my past experience prepared me for what happened next.

Kol came into me, body and soul.

Incredible sensations exploded all over my body as he took me down to the floor and covered me, easing between my thighs, penetrating me by inches as he watched my eyes. I could feel the power coiled inside him, now stretching and radiating as our bodies merged. I could taste his sweat on my lips, hear the delicious catch of his breath as our hipbones touched and he was deep inside me.

At the same time, I felt Kol inside my head.

His emotions poured into my mind, an endless cascade of need, wave upon wave of desire so deep and dark and powerful it swept away any conscious thought or plan I might have had. Then, like a net, it scooped me up and held me suspended, enveloping me in his heat and wanting, so much wanting that I thought I might scream from the pleasure of it. For in feeling his emotions, I discovered they matched my own so precisely we could have been reflections inside each other.

I should have murmured something, encouraged him, but there were no words. And he already knew what I was feeling, I could sense it, that I was as deeply inside his mind as he was in mine.

We became two streams of color and light and motion that met and sank into each other and grew stronger, bolder, more brilliant. On one level I could feel him loving me, my back arching as I met the thrusts of his body into mine, the sensations building and spreading within each of us. But Kol was behind my eyes, loving me there as well, enfolding me in the strength and safety of his emotions, reaching deep into my own for what he needed and wanted. And I gave him everything I had; I opened my soul up and took him in.

Being with him was beyond anything I'd imagined, and I rediscovered the man I loved in so many ways. In the gentleness of his hands as he stroked my breasts, in the deep laughter that came from him when I rolled him onto his back. He looked up at me as I moved on him, and looked down at himself from inside my head.

"Sajora." He sat up, wound his arms around me, and lifted me from the floor as he stood. The movement impaled me on him, and my head dropped back as my climax made me clench around him and shudder. "You honor me."

My back hit a wall, and I dug my fingers into his shoulders as his face hovered an inch above mine. His hips drove his shaft into me with heavy, powerful thrusts, and all I could do was hold on and watch the way his face changed, feel the tension rippling through his muscles. My own need returned with a vengeance, and demanded more than release.

I tangled one hand in his short hair, and brought his mouth down to graze against mine. "Come with me, Kol. Now, *now.*"

He buried himself inside me one final time, and said my name as his body jerked and shuddered. The moment I felt his semen jetting inside me, I came again, and the sound that ripped from my throat blended with his own deep groan of delight.

Slowly we slipped down the wall until we collapsed on the floor, our bodies still joined. I probably should have felt regret or guilt, but I couldn't. Not after taking him into me, where he belonged. Not ever again, no matter what happened.

Kol eased me on my side and cradled my face with his hand.

"I did not know it would be like this." He still sounded breathless and shaky. "I felt you inside my mind."

"You got in my head, too." I stroked the sweaty curve of his shoulder. "Is that supposed to happen?"

He smiled. "If it is not, we will not tell anyone."

I felt him getting hard inside me, and wound my arms around his neck. "Maybe we should check it out another time. You know. In case it was a first-time fluke."

He kissed the end of my nose. "Agreed."

We didn't sleep much that night, but it seemed unnecessary. Between making love and trying to shower together in the tiny lavatory, we talked. About everything, from Terra to Joren, the wrenching loneliness we'd felt, our dreams. It seemed as if everything we'd experienced up to that day at the warrior's quad on Joren was simply preparation for being together.

"We could make a life together, Sajora." Kol held me close and traced my features with one finger. "When we leave this place, we can find one that belongs to us."

"And the rest of the clan?" I tried to imagine what their reaction was going to be when they found out about us. "We can't just abandon them or ship them back to Joren."

"We will all go together. We were meant to be a family, although few may ever understand."

"Amen."

He met my gaze. "What about your sire, Kieran?"

I jumped a little. "You know?"

"I suspected, before we left Joren. It seemed too coincidental that he was Terran, and you wanted to divert his path. No, it is well, my heart," he added when I would have said something. "I understand why you did not wish to tell me, but I would never judge you for the sins of your sire. Do you still wish to pursue him?"

I was torn. On one hand, the filthy bastard had captured my mother and made her watch her friends be sold off to slavers, using her for his own pleasure. On the other, I had even more reason to live now, and Kieran was death to everything he touched.

I also knew someone had to put things right, or the seven of us would remain in a limbo, belonging to no world, no family. "I'm hunting him down as soon as we get out of here."

"Then we go with you." He rolled over and sank into me. "I will never leave you, Sajora."

I closed my eyes and sighed as his emotions poured into me. *I could get addicted to this real fast.* "Good, because you're stuck with me."

Several hours later I left him sleeping on my mat and went to meet Uel in the training room. I should have been exhausted, but nothing could dent the glow of satisfaction I felt. Kol and I were together, and nothing else mattered as much. There would be problems, of course, but now I knew we could work them out between us. We'd never be alone again, not for the rest of our lives. My romantic thoughts made me laugh a little.

This bond thing ought to be packaged and sold.

Uel was waiting when I entered the room, and asked me to secure the door panel. Only then did he remove his obek-la and gloves, and point to the exhibition platform.

"We will begin with countering the pivoting attacks within the *shahada.*" He drew out his blades. "Assume position on your mark."

"Good morning to you, too, Blade Master." I took out my tåns and shook my arms out before stepping up on the platform. "I'm already pretty good at pivot thrusts."

"I will make you better."

I slipped into the no-mind, no-self discipline almost immediately—that's how relaxed I was—and countered every move the Blade Master made. Until he slipped through a minute break in my guard, nicked my side, then called a halt to the match.

"I see what you mean by better." I looked down and touched the rip in my tunic. "How did you spot the chink in my guard?"

"You do not see; you feel the opening. As you feel part of the blades you use." He moved back a step. "Close your eyes." I did. "Now, tell me to which side of the platform I move."

I concentrated, listening for his steps. There were none, but I could feel a displacement of air to my right. "South."

"Correct. Open your eyes."

I looked down to see a knife at my throat. "Obviously I need to work on this feeling part, too."

"Obviously." He dropped his hand. "Now, return to your mark and repeat the exercise, but keep your eyes closed. Feel where my blade is, do not listen or look."

It took the rest of the session before I began to get the hang of anticipating his moves with my eyes shut. Before we finished the session, Uel turned off his visual accumulators.

"Attack me, and I will demonstrate what you will learn."

I attacked the now-blind Blade Master, but there was no move I could make that he couldn't anticipate. I circled around him, trying to figure out a new combination that would feign one pattern while executing another.

It took a lateral sweep, big enough to distract him from my other blade, which I pivoted and turned in on a quick side thrust toward his waist. As he moved to counter it, I reversed the sweep twice and ended up lodging my blade against his cranial case supports.

"Excellent. You learn quickly, Sajora." He switched on his accumulators. "Fayne is too confident in her own methods. She will not expect a sophisticated move like that from you."

I sheathed my blades as my happiness dimmed a few degrees. "I don't know if challenging Fayne is such a good idea."

"If you want to survive, you must." He pulled on his gloves. "You'd better report to your first session, before you are late. I will see you here tomorrow." He went to one of the wall panels and slipped out of the room.

I went to the door, but stopped when I saw a cluster of silvers at the observation window, looking in. In the middle of them stood Fayne, and she wasn't smiling.

CHAPTER NINETEEN

"Nothing is lost on the path that cannot be found on another."

—Tarek Varena, ClanJoren

I let the rumor circulate that I had broken off relations with my crossbreed family, and in session I made a point to stay away from them. Kol took it in stride—we'd talked about what I planned to do, that night we'd spent together—but the rest of the clan didn't like it.

"You should allow me to tell them why you have left us," he'd said, holding me against his chest just before we'd fallen asleep. "They will feel slighted and hurt."

"If you tell them, then no one will believe I've broken it off with you. Let them stew about it. It's the only way I can protect them." I traced a circle around his implant scar. "When this is over, then I'll explain everything."

Bek was kind enough to go along with my wishes, and matched me against other partners. With Uel's daily training, I swiftly improved, and began to prevail in every session. The trainer started bringing in other, more experienced trainees, but no one walked away with a win from sparring with me.

Weak, emotionally overwrought female that I was, I also stole a few moments here and there with Kol. Since claiming each other had done a lot to soothe the twin beasts riding us, there was no longer any strain in being together. Instead, the warrior-bond seemed to sustain us, even when we were apart. I felt a kind of confidence and security I'd never experienced in my life.

I even caught myself looking in the lavatory mirror and examining my reflection to see what Kol thought was so pretty. I had it that bad.

"I can't wait to get off this ice cube," I told him one afternoon as we ducked into a private corner. "Have you heard anything about the rescue vessels?"

He took me in his arms and touched his brow to mine. "No. Have you learned anything from Uel?"

"Only that he's getting very frustrated, waiting." I repeated some of the Blade Master's remarks, and added, "I'm sorry I can't tell the others why I'm treating them like they have a plague. It's for the best."

"You will not believe this, but Danea is more upset at you than anyone." He made a wry gesture. "Last night she accused Os of driving you away, and threatened to challenge him if he did not attempt to apologize to you."

I rolled my eyes. "Sparky misses me. Now I've heard everything." I patted his chest. "Do you miss me?"

"Since we were together, every night." He brushed his mouth over mine. "Although I am glad I no longer desire to tear the Tåna apart with my bare hands each time I see you."

Uel made some excuse to delay the Tåna-Shen, but each day that passed made the trainees and the staff more nervous. I sympathized with them—who wouldn't feel terrified, knowing that many ships were waiting to descend on the school?—but I felt confident that the Blade Master would get us out before some military genius started firing on the surface.

One morning Bek matched me against a second-level trainee who moved through the *shahada* with such hesitation and clumsiness that I gave up sparring with him after three moves.

"Take off your shades," I told him. "They're obscuring your visual field."

"No way," the trainee told me. "We've all been wearing them 'cause of that Skogaq. I'm not taking them off until we're off this planet."

I had noticed several of the second-levels had affected shades like the ones I'd just stopped wearing. I'd created a fashion trend that would piss off Fayne. Worked for me.

"Okay, look. Watch my arms." I showed him how to hold his guard against a reverse oblique attack without stepping away. "You have to maintain your stance, keep your weight even on both feet, and don't look down at them." He was smart to keep the shades on, but if Fayne or one of her buddies jumped him, he'd still be history. "What's your name?"

"Yen, and I'm never going to get this," he complained. "You're two feet taller than me, and twice as heavy, and *still* you move like you're made of air."

I didn't like the fact that he wore a brown band, and hadn't bathed since God only knew when, but I was trying to be more tolerant. All my experiments had proved successful with the other members of the clan; maybe I could do something with this kid—and convert him to a white in the process. I told him to take five and went to speak with Bek.

"I should not have matched him against you, but you have prevailed over the rest of the class." The Chakaran stroked his chin fur. "Very well, if you wish to augment his training, you may return after meal interval and practice."

"Thanks, Trainer." I went to tell Yen.

"That's very generous of you, Saj." He gave me a closed-lipped grimace that had to be his species' version of a grin. "I'll see you here later tonight."

Kol caught up with me in the galley and asked me about the trainee, and I filled him in.

"I do not like the idea of your being on the level after the trainers are gone," he said. "It is too good an opportunity for Fayne to attack."

I nodded and tossed him a near-apple. "Then meet me there, and watch my back. Ask some of the whites to come, too."

Kol finished eating before I did, and left the galley a few minutes later. I was in no hurry—I suspected Yen would prove to be a major pain in the butt—and I wanted the rest of the clan to leave before I headed out.

Sparky was waiting outside the galley for me, and wouldn't let me walk around her. "We will talk, Terran, if I must tackle you and sit on you."

"I left because I don't belong. As you've pointed out about ten thousand times," I said, trying to sound tough and indifferent. "Doesn't anything make you happy?"

"No." She folded her arms. "Not when I see Kol laughing and smiling, and you offering to tutor incompetents."

"Yeah, well, we're much happier apart." The urge to hug her made me bite the inside of my lip. "Can I go now? My incompetent student is waiting."

"Galena suffers in your absence." She lifted her chin. "And I have no one with whom to . . . share girl talk."

I couldn't help the burst of laughter. "That's priceless. Oh, Sparky, I do miss you."

Her expression became very serious. "Then come back to us, ClanSister."

I could have joked or said something to infuriate her. But the witch had just called me ClanSister, and I felt like hugging her again. "Danea, I . . . Give me a little more time, okay?"

She nodded and stalked off.

There were a group of whites waiting in the training room, although I didn't see Kol or Yen. "Hey, guys. Where are my men?"

One of the whites shrugged. "We've been waiting here since Bek opened the room. Kol went to Yen's quarters to get him."

"Yen didn't look too good today," another added. "Kol might have taken him to medical."

I suspected Yen was suffering from acute lack of confidence, but didn't comment—my first weeks on second level had been almost as bad.

"Where are his quarters?" Once I got directions, I nodded. "Okay, we'll scrap this for tonight. Thanks for waiting."

I went to the trainee's quarters to give Yen a piece of my mind and hopefully intercept Kol on the way. Maybe I could convince him to spend a couple of hours with me before he turned in for the night. As for Yen, he could wait until tomorrow.

I was surprised to see the trainee in a robe when he answered his door. "Are you feeling all right?"

"I've been talking to Kol; come in." He stood to one side, and as soon as I entered his rooms he shut the door panel and secured it.

There were a dozen silvers lounging around the room, but all of them stood as soon as they saw me. I turned with my blades in hand, in time to catch a solid punch to my face, and staggered backward. Four arms grabbed me from behind and disarmed me.

"She's rather pretty, for a hulking clod, isn't she?" Yen said in a very familiar voice. He shed his robe to reveal a small, muscular body the color of snow. Someone tossed him a towel, and he wiped his face, removing the thin layer of scales and smearing the dark pigment around his eyes.

I looked down at his penis, then up at his face. Her face. "Fayne?"

"And she's seen through my clever disguise." Fayne giggled and grabbed his/her crotch. "Does this bother you, Terran? I can make the bad male organ go away, like so"—he/she demonstrated by tucking his/her penis into a vaginal opening—"and I am female again. What do you think?"

"I think telling you to go fuck yourself is redundant." So Fayne was a hermaphrodite, and I was in big trouble. "Where's Kol?"

"In a moment. Hold her for me." Fayne came up and undulated against me, cupping my breasts and smiling up into my face. "I'm going to enjoy breaking you, Terran."

I lunged against the hands holding me, but someone smashed something down on the back of my head and the room abruptly darkened.

"Don't knock her out. I want her awake and watching." Fayne pulled on her trousers and tunic, and peeled off the dark wig covering her white hair. Whatever she'd used to glue it down had left dark brown streaks all over her.

My claws shot out, and someone behind me yelped and changed their grip. "How did you get the makeup and the wig?"

"I borrowed them from the original owner." She made a gesture, and one of the silvers opened the storage container. A body with a bloody, mutilated head fell out. The corpse's blood matched the brown streaks on Fayne's hair. "He didn't want to lend his face to me at first, but with a little sharp persuasion I was able to convince him to cooperate."

She'd skinned and scalped him, just to trap me. "You're going to die for this."

"Very good, Terran. You challenge me at last." Fayne made another gesture, and six more silvers emerged, dragging Kol out with them.

"Kol." I nearly broke free then. "No!" I whipped my head toward the albino. "Your fight is with me; let him go."

"I want him to watch, too. I am quite the exhibitionist." Fayne picked up one of Yen's chairs and smashed it on the floor, then took up one of the broken legs and hefted it like a club. "You know, Terran, I found some very interesting data in your medical files. You should have never let anyone put drone components into your body." She took a few experimental swings. "They're so . . . unreliable . . . under pressure."

Sweat broke out on my face. *Not my leg.* "You won't get a fair fight out of me if I'm crippled."

She smiled. "Why would I want a fair fight?"

I braced myself as she swung the club down, but nothing prepared me for the soul-shattering agony of the impact against the side of my knee. Through the sound of my own scream, I heard bone snap and alloy burst. A spray of metal bits, blood, and tissue splattered the floor.

"Unreliable *and* fragile." Fayne wiped some of my blood from the end of the club on my tunic, then took a second swing and smashed my knee from the other side.

I must have passed out for a few seconds from the pain. When I opened my eyes, I heard Kol shouting and saw bodies hurtling through the air around me. The silvers holding me dragged me back, but Fayne simply pulled out her tån and stabbed me in the chest with it. I writhed as the jolt sizzled through the white-hot pain of my shattered knee.

"That's one, Kol," she said, making him freeze.

"No, Fayne." Kol dropped the silver he was strangling and came toward her. "There is no need for this."

The albino stabbed me again. "And that's two."

He stopped. "What do you want?"

"A life." She glanced at my face, then his. "The problem is whose."

"Take mine," I said, panting through the pain. "My leg is ruined. You only have to stab me one more time. Easy."

"Sajora is finished; she will have to leave the Tåna." Kol wouldn't look at me. "Kill me and you will once again be champion, and will have vanquished both of us."

"God damn you, Kol!" Tears streamed down my face. "Shut up!"

He looked at me then. "Honor me, my heart."

"Your heart." Fayne sneered at me. "Look at her. Your heart is crippled, weeping, useless. Less than pathetic. I could have given you everything, you stupid male."

"Fayne, please." I would have gone down on both knees, but I couldn't feel my right leg anymore, and the hands holding me wouldn't let go. "Do you want me to beg? I will. I am. I beg you, kill me. Let him live."

"Thank you for deciding this for me, Terran." Fayne went over to Kol, who didn't twitch. "Shield me, Jorenian, and I will allow her to crawl out of here."

Vaguely I heard a click and realized someone had switched on a recorder.

"I, Jakol Varena, shield you, Fayne of Skogaq."

"Excellent." She stabbed him in the chest.

"Kol!" I screamed. "Run!"

More silvers came to hold him, but he didn't resist. He only spread out his arms and watched me as Fayne's blade struck him a second time.

"No, Kol, please," I sobbed.

Fayne's laughter and the voices of the silvers dwindled away for a moment as our eyes met, and I heard only Kol's voice.

"I will be waiting for you," he said, in a voice so beautiful and tender that it caressed me like an actual touch. "Walk within beauty always, my heart."

Then Fayne stabbed the man I loved a third time, and I watched him die.

I remembered very little from the time I saw Fayne kill Kol to the moment Uel's voice made me open my eyes. Someone clubbing me on the back of the head again. Being thrown against something hard. Crawling, trying to get away. And around everything, the mind-numbing horror of my shattered leg, and my broken heart.

"Sajora." Gentle hands eased me over onto my back. "Be still; you are injured."

"Fayne." My voice had gone from screaming. I looked down the corridor and saw a long, twisted trail of blood. My right trouser leg was sodden, wet and red. "Fayne killed Kol."

"I know. We are taking you to the infirmary now."

I didn't want to go to medical; I wanted to be with Kol. I groped for my blades, but they were gone. I struck out at the blurred face over me. "Let me die!"

The Blade Master didn't even try to dodge my fist. "Then Fayne will prevail over both of you. Is that what you wish, Sajora?"

That question haunted me through the next days, as the medical people tried once more to patch my knee back together. Most of it I spent drugged, and when I did have a lucid moment, I usually saw some nurse or doctor standing beside my berth, making notes or checking my vitals.

"I want to speak to the Blade Master," I said, croaking the words until a nurse brought me some water.

"He will be here tonight to visit you." She did something to the infuser line in my arm, and my eyelids grew heavy. "Rest now, while you can."

Uel appeared the next time I gained consciousness. He drew back the linens to look at my leg. I felt a vague sense of surprise to see it still attached. "The doctor has not removed the ruined components. I ordered him to wait until you have decided what you wish to do."

"There's no decision to make. I knew what the risks were." Everything felt dull and faraway. "It has to be amputated."

"Can you tell me what happened?"

I focused on his artificial face. "It was a trap. Fayne and her silver buddies were waiting for me. She smashed my knee and killed Kol."

"The Skogaq says you killed Yen to lure her into an ambush," he said. "She admits to being responsible for your injuries and Kol's death, but claims inflicting them was self-defense."

I'd existed in a universe of pain up until that moment. Now my inner beast clawed through the suffering, tearing everything good and kind within me to shreds. "She's lying."

"She has witnesses." He put a gloved hand on my shoulder. "You do not."

I slumped back against the pillow. Of course, she could get the silvers to say anything she wanted. "So I'm out of training, and she wins. Just like that."

"The League and the Hsktskt have ordered me to conduct the Tåna-Shen tomorrow." He covered my leg. "I must, for my allies will not reach us in time. Once the trainees are dispersed, I will be allowed to remove a small contingency of staff from Reytalon in my personal vessel. I want you to come with us, Sajora. There is nothing left for you here."

"I don't care what happens."

"I do." He touched my hand. "It is for the best. Try to sleep now."

After he left, I didn't sleep. I replayed the scene in Yen's quarters a thousand times, late into the night. I should have been able to save Kol; I should have gotten free of the silvers and cut Fayne's throat. I couldn't move my right leg without biting back a shriek, but my hands still itched for my blades.

She dies for this. Somehow, I have to find a way.

"Thinking about killing, are you?" Bek circled around my berth and studied my face before handing me my tån and my Omorr blade. "I thought you'd feel better if you had these."

Nothing felt as good as those hilts in my hands. "Thanks, Trainer."

"You are welcome." He pulled up a chair and sat down beside me. "Uel told us you weren't coming back. Forgive me, but I had to see that for myself."

The Chakaran's brusque affection might have touched me, if I'd had a heart left in my chest. "Well, you've seen me."

"Your father suffered something similar to this. I thought it would break him, but he prevailed in the end." The trainer cocked his scarred head. "A pity his daughter hasn't the same spine."

I couldn't believe it. He was comparing me to Kieran, when my life had just been destroyed and the man I loved sacrificed for nothing? I drove my tån into the berth mattress. "My leg is *gone*. So is Kol."

"And he took your courage with him."

"What do you want me to do, Bek? Limp out into the arena and challenge the fucking little bitch? I can't even stand up straight."

"I've never heard you whine before. You do it well." He showed me his teeth. "Have you tried standing up straight?"

I looked down at the traction rig webbing. "No."

"You might like it better than whining." He rose to his feet, then showed me something else—an infuser instrument the nurses had used to administer my drugs. "I've calibrated this for your weight and physiology. It will deliver enough localized neuroparalyzer to deaden half a limb. You infuse yourself behind the knee. Or what's left of it."

"Even if I do, they won't let me out of here. They've secured the door panel."

"Memorize this code." Bek gave me a series of numbers, then waited for me to repeat them. "That is the Blade Master's access code. It will open any door in the Tåna."

"Why are you doing this?"

"I once considered your father my finest student. Until you." He tucked the infuser under my linens. "Now I must go. I wish you luck, daughter of Kieran."

I didn't think very long or hard about what I was going to do. I watched myself pull the intravenous needles out of my veins almost at a distance, as if it weren't happening to me. The traction rig took longer, until I gave up trying to release the straps and used my blade to cut my leg free. Standing took twenty minutes, and my medications wore off by the time my right foot touched the floor.

Pain filled the world.

"Are you suicidal?" the night-shift doctor shouted when he saw me upright. He tossed down a chart and came racing toward me. "Get back in that berth. Nurse! Assist me with this patient!"

I adjusted my hold on the bed rail and waited until he reached for me, then poised a blade at his belly. "Step away."

"Your knee is crushed. You cannot walk." He looked down and paled. "You have started bleeding again."

"Stop the bleeding and splint the knee."

"I will not." He winced as the tip of my Omorr blade penetrated his physician's tunic. "Even if I stop the bleeding, you'll pass out from the pain with the first step you take."

"Then you can throw me back in the berth." I bent down and applied the infuser with Bek's neuroparalyzer, then straightened. "Do it."

I stood and fought waves of dizziness as he knelt, unwrapped the bandages, and cauterized two places spurting blood. Looking down, I felt a little better—my upper and lower leg bones were still intact. Only my knee joint had been demolished.

After my leg was splinted and rebandaged, I ordered a nurse to bring me garments and help me dress. The doctor stood waiting for me to pass out, and got more irritated by the moment when I didn't.

"Whatever you've infused yourself with will wear off," he told me. "Even without the pain, your leg will not tolerate your own weight. Now that there is nothing left of your knee, the ends of the bones will grind together, then snap."

"Give me that suture laser." When he didn't move, I sighed and grabbed the nurse, pinning her against me with one arm and holding my blade to her throat. "Now, *give* me that *suture* laser."

He handed it to me. "You won't get far."

I picked up the syrinpress they'd used to sedate me, and infused the nurse. "You'd be surprised, Doc." I let her fall on the berth, and grabbed the doctor as he rushed forward and gave him a dose. "Pleasant dreams."

I held on to anything I could and dragged my leg the first hundred yards, until it became apparent I needed a crutch. Transmuting my tån to long-sword form, I drove the tip into the floor and leaned on it as I pulled my numb leg into my first step. Nothing snapped, nothing popped, but I did hear metal grind on metal, and a liquid, mushy sound.

The ends of my bones. The shreds of my muscles.

I had three choices of destination, but I knew who I had to see first. There was only one person who could help me get Fayne alone, so I went to the Blade Master's quarters. The code Bek had given me worked like a charm, but when I let myself in I found his rooms empty.

Maybe reconstructs are like Ren, and don't have to sleep.

I was afraid to sit down to wait, and kept pacing around the spartan front room, wondering if I should try to access his console and signal him.

The drugs numbing my leg made me feel sluggish and cold, and I opened his garment storage unit, hoping to find a cloak or jacket I could wear. Instead I found a second console, built into the unit. According to the screen, it was maintaining a direct, open relay to some point outside Reytalon.

Has to be the ships he's been trying to get through the rift.

Since it had been left open, I could access the database and replay the last transmission, see what was happening with the rescue effort. I pulled up the archives and saw messages had been going out every few hours.

Uel must feel so frustrated. Then I saw a very familiar personal relay code. *That can't be right.*

I keyed up the last relay sent to that particular code and put it on vid and audio. The wolfish grin of my old friend Thgill filled the screen.

"You worry too much, old man," 'Gill said. "What is it now?"

Although Uel's synthetic face didn't appear on the screen, the audio playback responded with his voice. "She is in a great deal of pain, Major. You're sure you can rebuild her leg?"

"Pain is to be expected. I wouldn't have told you to have your Skogaq hobble her if I couldn't deal with it." 'Gill leaned forward. "My commander wasn't happy to hear you've made a deal with the Hsktskt, you know. Even if you are giving us half of your trainees, he thinks it's treason."

"Let him think whatever he wants, as long as I get my money." Uel made a sound approximating a laugh. "You're coming down on-planet for the Tåna-Shen tomorrow, aren't you?"

"Now that Jory's stuck in medical, there's no reason for me to stay away. I'll bring her new leg with me, so you can have a look."

"Keep the leg. We can't do anything until we move her off Reytalon. Bring my payment."

"Whatever you say, boss."

The audio and vid switched off.

I removed any sign that I had been in the Blade Master's quarters, then limped out into the corridor. The computer room Kol had shown me was only a short distance away, and I used Ucl's access code to let myself in and secured the door. Then I changed the code, so no one but me could enter or leave.

Bring my payment.

Whatever you say, boss.

I accessed the Tåna database, and pulled up a number of files. It was all there—names, dates, transaction fees. The order was nothing more than a glorified slave ring. They trained their students to be highly skilled assassins; that much was true—then put them on the block and sold them like cattle.

I turned off the vid and concentrated on unraveling the tangle of lies and deceit that had brought me to this place, this moment. Of course, in hindsight, it was deeply suspicious that a person who could fix the illegal tech holding my knee together just happened to be on the Terran trader. And just happened to save me from two of his men. And just happened to strike up a friendship with me.

Thgill, waiting somewhere up in space with a new leg for me. Why? Why would Uel hire him to fix my knee?

At the Rilken outpost, I'd bought passage on the *Chraeser,* but now I recalled several other stewards turning me down flat, before I'd even made an offer. Had they been paid off? Probably. And the *Chraeser*'s steward had actually approached me first, claiming he'd heard I needed passage out of the system.

Which brought me to the Shadow—he must have been Uel. As a reconstruct, he could change his voice, height, and weight simply by adjusting a few controls and switching out some components. And the dimsilk garments he'd worn had effectively camouflaged his form.

The Shadow tests me—auditions me—then plants the idea of becoming a blade dancer in my head.

He couldn't have done anything on Joren; I'd made that mess myself. But he must have sent Uzlac to me. The Ramothorran had also approached me—not the other way around.

But why would Uel want me on Reytalon? Why have Fayne cripple me? He could have killed me hundreds of times, anywhere along my journey. Why keep me alive?

Is it just another kind of audition, to find out if I'm cold-blooded enough to be like my father? I thought for a moment, then turned cold down to the bottom of my soul.

It wasn't an audition.

You learn quickly, Sajora.

He was training me.

. . . you will become what you were meant to be.

Training me to become like my father.

I copied everything about the enslavement of blade dancers from the database onto disc, sent three tersely worded signals, then limped out of the computer room. It took me a long time to get to my old quarters, and from the sound of my knee I didn't have much time left.

I didn't care. Fayne had to pay for what she'd done to Kol and me. And Uel; I had special plans for him.

Everyone was sitting around sulking as I staggered through the door. Only Galena got up and started toward me, but Osrea put a hand on her arm to stop her. No one looked directly at me.

The Jorenian form of the cold shoulder, I'd assume.

"Hi, Jory," I said, deliberately loud. "Good to see you, how's the knee, we were so worried about you. They had to beat us back to keep us out of medical."

"We were—" Galena started to say, but Danea rose and stood between us.

"Kol embraced the stars for you, Terran." Her hair seethed like wild yellow snakes.

"I know; I was there." I scanned the other faces. "Why aren't you celebrating? Shame to waste such a great opportunity to party."

Sparky's fists clenched. "I should kill you now."

"You'll have to get in line," I said, and limped over to the garment storage unit. "I've challenged Fayne."

CHAPTER TWENTY

"CHOOSE YOUR DIRECTION WISELY."

—TAREK VARENA, CLANJOREN

"What say you?" Nalek shot to his feet. "Have you lost your senses, ClanSister?"

"Something like that." I pulled out one of Kol's tunics and held it against my chest. A little big, but it would have to do. I balanced myself on my good leg while I stripped off my own tunic. The neuroparalyzer was wearing off; I could feel a dull, fuzzy throbbing gathering around my knee. "I'm finally doing something a real Jorenian does."

"Committing ritual suicide?" Os asked.

"No." I slipped on Kol's tunic and turned to face them. "I declare ClanKill on the Skogaq Fayne."

Sparky nodded. "She will do the work for me, then."

"There's something else you should know. Uel hasn't done anything to get us out of here. He's auctioning us—all of us—to the League and the Hsktskt, immediately after the Tåna-Shen. Some of them are coming down to watch the bouts, see who they want to buy."

Galena went white. "But he promised he would help us escape the war."

"He lied. People do that." I limped over to Renor, feeling the vague pain begin to knot and tighten. "You told me you worked repairing drones for your HouseClan, right?" He nodded, and I reached down and tore my trouser leg open. "Can you fuse this enough for me to stand and walk without a crutch?"

The bloody mess that had been my artificial knee seemed

to shock everyone but Plas-Face, who crouched down to examine it.

"Would not the physician do the work?" he asked me.

"No. He says it has to be amputated. What I need is for you to stabilize these"—I showed him the ruined brackets—"enough to bear my weight for a couple hours."

"That will seat the broken ends of the bones against each other." He gave me a dispassionate look. "The pain will be unbearable."

I pulled the infuser out of my trouser pocket and waggled it. "I have drugs."

He nodded. "I will need something to melt the alloy."

"Here." I put a hand on his shoulder as I lowered myself to a mat, then pulled the suture laser I'd stolen from medical out of my tunic. "Do it."

As Ren adjusted the beam and went to work, the others slowly formed a ring to watch—all but Danea, who made an effort to look like she didn't care what he did. I hid a smirk as I caught her taking peeks anyway.

The brackets conducted the heat from the laser into my flesh, replacing the numbness with fresh pain. I took hold of the mat with both hands. By the time he'd finished one side, sweat poured freely down my face, and I had ripped out two handfuls of matting.

"Stop," Galena said as she knelt beside me. "Please. Please stop doing this to yourself, ClanSister."

Renor paused and glanced up at me.

It would have been easy to give in. Everyone certainly expected me to. I could signal someone, have them come and take me back to that comfortable berth in medical, and—once they woke up the doctor and the nurse—get some more of those great painkillers the doc had given me. I'd been through enough, and I definitely deserved a break.

I will be waiting for you.

I avoided Birdie's touch and set my jaw. "Do the other side, Ren."

He bent his head and went back to work. The pain went past

unbearable soon after that, and I breathed through waves of it, like a mother giving birth. Though it burned, the laser's heat effectively cauterized the last of the raw tissue. Welling blood from the stress of walking slowly cooked until it dried crackling and black. Deep inside, I could feel the splintered ends of my bones drawing together.

To keep my focus, I concentrated on the faces around us. "I'll need you all there to watch my back and keep Fayne's buddies out of the quad, in case she's told them to make sure she wins."

Nalek opened his mouth, then closed it and averted his gaze.

Osrea wasn't so diplomatic. "We cannot go with you."

"What are you talking about?" I looked from him to Danea and Birdie. "This is ClanKill. This is for Kol. Of course you're going."

"Kol shielded her, Jory." Nal sounded miserable. "We cannot act against her."

"She forced him to shield her in exchange for my life." I saw their eyes and let out a slow breath. "You can't be serious. She murdered Kol. *She killed our ClanBrother.*"

"We are not a HouseClan," Sparky snarled at me. "We never have been. Now that Kol is dead, we have a responsibility to a real HouseClan—Varena. We will respect his shield and find a way to return him to his House, where his embrace can be properly celebrated."

I sputtered a laugh. "You think Uel is just going to let you have his body and give you a ship to jaunt back to Joren?"

"We will find a way."

"You know, Sparky, I don't like you. I've never liked you. But I never thought of you as a coward." I eyed her hair. "No wonder it's yellow."

She got down on my level and in my face. "You I should kill. You who diverted his path, and delivered him into Fayne's hands. You who stood and watched and did nothing while he was murdered. You who dare call me craven."

I could have corrected her, told her about Yen and how Fayne had tricked me. Somehow I just wasn't interested in defending myself to her. "You keep thinking that way, Sparky, if it makes you happy. When I'm done with Fayne, you and I can dance."

"You will not be dancing." Ren turned off the laser. "You will be able to stand for a short period of time, but eventually the pain will overwhelm you, or the tibia will split. Your leg will collapse."

"Whatever." I used my sword to get to my feet, then tested my weight on the leg. The fused metal groaned, but held. The pain made my vision blur, but I was going to save the drugs for the arena. "Thanks."

Ren brought me a new pair of trousers and helped me get them on. No one said a word until I headed for the door panel.

"I said a foolish thing to you once, ClanSister." Birdie moved until she was blocking my exit. Tears ran freely down her face. "I said nothing hurts you. I was wrong."

"Yeah, sweetheart, you were." I reached out and wiped her cheeks with my fingertips. "Stay here, okay?"

Renor hovered on my other side. "Fayne will kill you in the arena. You cannot beat her in your condition. Let me take you to medical, Jory. Save your own life."

"My life ended with Kol's." I looked down at my ruined knee, then at Danea's brooding expression. "I loved— I honored Kol Varena. On Terra, honor is more than what you feel toward another person. It's personal. It's what's inside you that makes life, pain, and even death bearable. Honor won't be satisfied—fuck, I won't be satisfied until I get some justice for Kol's murder. If it means I have to die, then I will. It's a fair trade, Ren."

I limped out and headed for the quad alone.

I knew from the moment I'd sent the challenge signal to Fayne that I would die in the quad. Knowing the only real family I'd ever had wouldn't stand with me didn't matter. It made me a little sad, but in a distant, disconnected way. Besides, I didn't want them to die at the hands of Fayne's silvers.

This way was for the best.

Other trainees in the corridor stepped out of my way as I

dragged my mostly useless leg and hobbled toward the third-level entrance. Some of the silver bands laughed and pointed, but a number of browns and whites stood stiff against the wall panels as I passed them. A few gave me a brief salute with their tåns.

All hail the limping hero.

At last I got to the end of the corridor. I heard steps behind me, and saw that a small army of trainees was shadowing me. It reminded me of how fans used to dog me during pregame warm-ups at the arena. Well, I'd definitely be giving them a good show today. I paused to wipe the sweat from my face, then entered the third level.

The Tåna-Shen had already begun, judging by the rainbow of bands waiting to enter the arena. Someone shouted as I came in, and the milling crowd parted in front of me, forming a tightly packed gauntlet—silvers on one side, browns on the other. No sign of our whites. I ignored them as I limped toward the now-empty quad.

Unlike the time I'd been given a spit-bath on Terra, no one had anything to say. Slowly tåns began to appear, and the browns started slapping them together in time with my halting footsteps. The silvers kept their blades sheathed, but as the noise grew louder they began exchanging worried looks.

Dursano got in front of me, but someone dragged him aside. I met Bek's calm gaze and nodded my thanks. The Chakaran inclined his head.

The clack of the tåns grew deafening as I mounted the platform. Fayne appeared on the other side of the quad, completely outfitted in white. Someone offered me an obek-la, but I shook my head.

Let her watch my face. Let it be the last thing she ever sees.

"You look terrible, Terran," she called out to me. "Whatever have you done to your leg?"

"A rat bit me." As the browns around me snickered, I rolled back the sleeves of Kol's tunic. "A little diseased white rat."

"What a pity." She grabbed the ryata and flipped up and over the top cord, landing on her feet with a flourish of spinning blades. "Can you still dance?"

I lifted my gaze to meet hers. "Oh, yes, Fayne. I can dance."

As I prepared to climb into the quad, someone reached out and plucked me backward.

"Hey." I turned to face a blade dancer in full dimsilk. "You're making my rat wait."

"Your pardon, lady." A strong blue hand pushed back the obek-la, and a ghost smiled down at me. "I believe I have prior claim."

I nearly keeled over. Then I did the second-worst thing and dropped my blades so I could touch his face. "You're alive."

Kol touched his brow to mine. "I live, my heart."

"But I watched you die." I couldn't stop touching him, running my fingers over his hair, gripping his shoulders, pressing my palm against his chest—where I made an interesting discovery. "Your implant's gone."

"They removed it. There is no poison in it, only a substance that causes one to mimic death." He lifted my hand to his mouth. "They took me to a processing center some distance from here. Those who fail the Tåna have their memories wiped clean of this place; then they are sold as slaves."

"But you're here. You remember me." I knew I sounded like an idiot, but it didn't make sense. Kol, alive, holding me in his arms. Not dead, not mindless, not a slave.

"Bek persuaded them to set me free before they removed my memory."

The Trainer must have known what would happen to Kol—but why would he help us? Was he a slave, too? "Didn't anyone try to stop him or report this to Uel?"

"By that time, there was no one left alive to do so. Bek had to help me remove the implant." He looked around the arena. "The others must be told. Why are you not in medical?"

That reminded me. "I have something I have to do." I kissed him. "I love you, Kol."

"Permit me to do this first." He took my hands in his. "I Choose you, Sajora Raska."

My eyes went wide. "You do?"

"I knew from the moment I first saw you. There is no

other; there can be no other." Then he hooked his leg around the only one I had left working, and knocked me off balance. Before I could grab on to something, he hefted me up in his arms and tossed me to the side, where three whites caught me and held me.

"Damn it, Kol!" I was crying as I watched him climb into the quad. "You stay alive!"

All around us, Fayne's silvers moved in, enclosing the browns, cutting off the quad until they formed a living wall. There would be no way we could escape, even if Kol defeated the albino.

I limped up to Kol's corner, where white neutrals were assembling en masse. "Let me up front."

Many hands helped me until I was on the platform, with a clear view of the quad. In the center, Fayne and Kol were facing each other, blades out in raen-tån form, waiting for the other to move.

The Skogaq looked a little paler than usual. "I killed you once," she muttered, swinging her sword in a fancy weaving pattern, while Kol simply stood and watched her. "I will cut off your head to make sure you stay that way this time."

He lifted a hand to show her his claws. "I declare you my ClanKill, Fayne of Skogaq."

They moved in toward each other, and the crowd erupted with shouts and cries as the two swords crashed together. The albino whirled and countered with a short thrust, reversed a heartbeat later but met by Kol's blade. From the look on his face I knew he had already become part of his sword.

Stay alive, I prayed. *Just stay alive.*

They fought with the long swords so ferociously that sparks flew, with moves so fast that I could hardly follow the patterns. Fayne worked around him, jabbing feints at his guard, trying to find an opening. Kol remained centered, letting her whirl and dive, conserving his energy while counterattacking with cool, calm precision.

On a turn, Fayne abruptly transmuted her blade out of Kol's sight and whipped around to use the thion-tån form to move in and get under his longer sword. Kol knocked her back with a well-placed elbow and matched her blade length. The distance between them as they fought began to close.

I finally understood why they called it the dance as I watched my lover and my enemy battle. The patterns and movements seemed accompanied by music only they could hear—terrible, beautiful music that echoed in the whisper of their breath, in the crescendo of the blades meeting. Voices quieted as everyone in the arena fell under the spell of seeing the savage grace, the clever moves, that should have left dismembered limbs all over the quad.

Fayne liked to use transmutation as part of her attack, hiding the fact that she had shortened her blade during a turn or a reverse, but long before she had reached elok-tån form, Kol had already begun to anticipate it. I could see the frustration building in her; it etched her narrow face with faint lines and mottled patches of hectic color.

Then she did something no one expected—Fayne retreated, backing away toward Kol's corner. The browns hooted, but the silvers around us seemed to be smirking. I saw why when she turned and thrust her elok-tån under her tunic, then brought it out in osu form. She used both blades to slash at me, leaving her back vulnerable to Kol's attack.

She must be more desperate than I thought. I easily stepped out of reach, then went still as I saw her pivot on Kol, who had followed her across the quad. "Kol, watch out!"

The sound of her tån burying itself in his arm wasn't right, and I saw why when she jerked it free.

Her holographite blades wouldn't dematerialize—Kol's implant, which controlled that, was gone.

Kol parried her follow-through thrust and spun around, only inches away from me. His blades were reacting to her implant, so they would not stay solid. I had the feeling he meant to use his claws on her—Jorenians used no weapons to carry out ClanKill. He needed an edge to get there, so I jerked the Omorr blade from my belt.

"Kol." I leaned in and used two seconds during one of Fayne's fancy patterns to toss him the knife. "Use this."

Fayne saw what I'd done, and her face turned pink. "Kill them all!" she shouted.

That was when the silvers all around the arena pulled out their blades, and began attacking the browns and the whites.

I couldn't watch the bout any longer; I was too busy countering attacks from two directions. Three of the whites around me fell quickly, and although I knew they weren't dead, it didn't help.

"I should have known the little bitch would turn this into a wholesale slaughter." I grunted and shoved a silver back onto another white's blade. From the frantic clash of blades behind me, I knew Kol was holding his own. But with the deep wound in his arm, how much longer could he do that?

A huge, block-bodied silver launched himself at me, only to be brought up short. He gave me a peculiar look, as if it were my fault.

"It appears"—someone thrust the body away from me—"you need a little help, ClanSister."

I had never been so glad to see Snake Boy's ugly face. "Hey, Os. Join the party." I parried a side blow from another attacker, and braced myself against the corner of the quad as I slashed my tån across the furry being's eyes.

"I believe I was invited, too." Nalek flanked me on the other side. "We should move away from the quad."

"No. As long as Kol is in there, I'm staying right here."

"Kol?" Nal tilted his head before punching a silver out with one huge fist. "Kol is dead."

I'd thought they'd come because they'd heard the news. "You came here just to help me?"

Os rolled his eyes. "We wanted to hurt someone. You were our excuse. What say you about Kol?"

Nal glanced up into the quad. "Mother of All Houses, he lives." He grabbed a silver around the neck with one big arm and squeezed it like a vise. "You hear that, scum? My Clan-Brother lives!"

Something swooped down at me, and I barely parried a vicious sweep at my face before it took off. The move cost me, and I almost slid to the floor before grabbing the edge of the quad.

"Jory, back-to-back." Os held out his lower limbs, and I dragged my awkward form behind him. His double joints allowed him to reach back and grab me around the waist, anchoring my back against his. "Our ClanSister and her avian friends have arrived."

I looked up to see Galena and a handful of white avatars sailing overhead, countering the attacks of the silver avians. "She's going to get herself killed."

"She is tougher than you think," Os said, and grinned as our little Birdie drove a silver twice her size into the dome wall, breaking one of his wings and sending him spiraling down.

We couldn't watch her. The silvers were bunching around us, kicking aside the bodies of their fallen comrades in an attempt to get at the three of us.

"There are too many," I heard Nal mutter. "We must retreat."

"Go." I tried to untangle myself from Snake Boy's backward grip. "You two go, get Birdie, and get out of here."

Os turned his head until his cheek touched my hair. "We are HouseClan, you stubborn female. We stand together, we live together, and we will die together."

Fayne's clique nearly overwhelmed us with their sheer numbers, until a strange glow appeared over the arena, and everyone stopped fighting for a moment.

I saw Danea standing at one end of the arena, her hair standing straight up, her arms open wide, as if she wanted to give everyone a group hug. Something told me to look at the other side of the arena, and there was Renor, glittering and silent, absolutely motionless.

They're combining their fields.

The glow descended like a soft, golden cloud, crackling a little as it enclosed the quad, forming a circle that drove back everyone—even our white comrades—on the outside of it. The boys and I stood on the inside of the glow, which didn't touch us, right beside the quad. Then the barrier began to swirl and move, like a stream of liquid light, growing brighter, whipping around the quad like a perfectly circular river. I heard screams and curses as some of the silvers tried to hurl themselves through it and were thrown back by larger, crackling jolts of energy.

"Remind me to apologize to Sparky later," I said to Os.

The bioelectric field maintained its protective circle around us, and Danea and Renor showed no signs of tiring. A couple of silvers tried to attack them, but bounced off an invisible wall whenever they got within a meter.

I could watch what was happening in the quad now, and when I turned I saw Fayne was visibly tiring. Kol's strategy of remaining centered and letting her exhaust herself with her spinning attacks had worked.

The Skogaq must have known she was beaten, for she lunged in and tried to spit in Kol's face. He turned his head and slashed at her, knocking one of her blades out of her hand.

Disarm.

Kol's follow-through left a long, deep diagonal gash across her mouth and shattered her front teeth. Her pale blood poured down the front of her dimsilk as she grabbed her face and cursed him through her ruined mouth.

Disable.

The wound compelled her to a final, horrifying attack, during which her limbs whipped and thrashed around Kol in a frenzy. He countered each thrust, matching her speed, his eyes following her patterns and catching each blow before it landed. At last she faltered, creating a tiny opening in her guard. Kol must have been waiting for it, for he thrust my dagger in and stabbed her in the side.

Fayne staggered back, holding her abdomen, her eyes wide with disbelief. Then she sank to her knees, struggling for breath.

He punctured her lung.

Kol dropped his blades and advanced on her. His claws gleamed dark blue as he planted a foot on her chest and knocked her flat on her back.

Outside the protective circle, the battling silvers and browns and whites stopped fighting and turned to watch the quad.

My beloved leaned down. "My people eviscerate their enemies. Slowly."

"Do it!" she said through her mutilated lips.

"I will not sully my hands with your filth. You are nothing but a pawn, Fayne." He looked up as Uel entered the quad. "There are better dancers, and more important kills."

"I will see to the Skogaq. You have prevailed, Kol Varena." Uel stepped to one side and gestured for him to leave the quad.

As Kol moved toward his corner, coming to me at last, Fayne rolled to her side and grabbed one of her blades.

"Behind you!" I shouted.

"No one laughs at me!" she shrieked, and drew her arm back to throw the blade at Kol's back. Then she looked down and choked as Uel's blade buried itself in her chest. A stream of darker blood dribbled from her mouth; then she fell over and went still.

Dispatch.

Trainers and inductors broke up the last of the fighting, and the injured were taken to medical. Kol and I and the rest of the clan spread the word among the trainees about the truth behind the order, and gathered the survivors in the galley.

No one challenged us. It seemed now that we'd destroyed the facade of the Tåna, everything fell apart. Also, I doubted any of the staff wanted to try to force the issue. We outnumbered them twenty to one, and keeping us in the dark had been their only method of control.

"I'm not concerned about the Tåna slavers. The Hsktskt and the League are still out there," one of the browns said as he wrenched the band from his arm and tore it in two. "They're going to try to take us, whether they pay the Blade Master or not."

Time for my little announcement.

"Listen up!" I waited until I had some quiet. "I've sent two signals, one to the Hsktskt flagship, the other to the League. I told them we've staged a rebellion, and they can forget about the deal with Uel."

Several voices rose in disbelief, agreement, and anger; then Kol stood up and gestured for silence.

"Our training here at the Tåna has well prepared us for battle. The difference now is, we know why." He looked around the room. "For want of a better word, this is our territory. Either we surrender it and become slaves, or we take back our freedom."

"Kol Varena." Bek appeared, and stood unresisting when two whites grabbed him. "I have news you should hear."

Kol gestured for the trainees to release him.

"What news could we want from a slaver?" one of the neutrals demanded.

The Chakaran waved a paw. "I am not a slaver. I, too, joined the order for the same reasons the rest of you did. Uel kept me to serve the Tåna."

"And you had no problem training the rest of us to be good little slaves." I shook my head. He might have saved Kol from being mind-wiped, but I still didn't trust him. "Pull the other one, Bek."

"It also helps to have a cardiac implant." He opened his tunic and showed us the scar over his heart. "When you graduate from the Tåna, the neuroparalyzer is replaced with real poison. The trigger is sold, along with you, to your owner. Mine is Uel." He smiled a little. "I expect my death will be quite soon now, but until then, I can be of help to you."

Blade dancers know two fundamental truths. The first is that they will die. The second is to live each moment as if death awaits them in the next.

I gestured toward Kol. "Why take the chance to save Kol, if you knew Uel would kill you for it?"

"I saw an opportunity to end all enslavement by the Tåna. That was worth my life." He pointed to the observation viewer. "Ground forces from both sides have begun landing on the surface. They will invade the Tåna within the hour, and I am certain neither side intends to let the other have you. Uel and the rest of the staff have already taken refuge in the bunkers on level one."

"We are not going to hide or surrender," I told him. "We're going to do what we were trained to do, and fight."

CHAPTER TWENTY-ONE

"THE PATH IS YOUR ONLY HOME."

—TAREK VARENA, CLANJOREN

Kol decided to trust Bek, but assigned two whites to watch him as they gathered up the remaining trainees from second and third level. He then created a command post in the computer room and brought up plans for the entire Tåna facility.

The school itself resembled a beehive, with its stacked levels and network of rooms and corridors. Not surprisingly, there were only two entrances and exits—one leading into the first level, which Bek assured us Uel and his people had already sealed, and a passage that began in medical and led out to the surface, where failed trainees were transported to the separate processing center.

"They can just blast their way in through the dome, can't they?" one of the former browns asked.

"That would destroy the interior environment and kill all of us," I reminded him. "They're pissed, but they like their slaves alive."

"According to this, they will enter through the medical facility. We must evacuate all the wounded to a safer area"—Kol pointed to a remote training room on the second level—"here."

"Initially, the Hsktskt and League will fight to gain control of the access tunnel," Bek agreed. "That will give us time to set up a welcoming committee in the trainee quarters section."

I frowned at the schematic. "But we can't hold them off, once they get inside. We don't have their firepower. Plus there are four entrances and exits from medical into the facility, and they can blast as many more holes in the interior walls as they like."

"True, which is why those positioned in the corridor will only draw their fire to the third-level arena." The Chakaran traced a path on the schematic. "Once you have the soldiers in the open, you will have room to attack."

"There is not a great deal of cover." Kol studied the screen. "Yet I believe we can create some advantage."

"We can." My shockball training kicked in, and I outlined a defense plan that would have sent my offcoach into a frenzy of delight. "It only works if the flanks move in tandem. We're going to need some kind of headgear we can use for signaling."

"I can provide those from the trainer supply room," Bek said.

"Wait." I took out my Omorr blade. "We need you, Bek. We need to get that implant out of you."

"The poison is triggered by removal, as well." Bek looked thoughtful. "If Uel had wanted me dead, he would have killed me by now. He must be keeping me alive for some reason."

"Like what?"

The Chakaran touched his scarred face. "I saved his life once. Perhaps he is returning the favor at last."

Kol frowned. "We need to provide you—and all of us—with some protective cover. How much dimsilk is available?"

"There is not a great deal stored here."

"Get as many people as you can in dimsilk; the rest of us will wear black over insulating wraps—and shut down the lights," I said. "The dimsilk and the wraps will keep their thermal gear from detecting us, and what they can't see, they can't shoot."

Os leaned in. "We can drop the temperature. The cold will make the Hsktskt slower to react."

"All cold-blooded trainees will have to wear thermals, then, or the same will happen to them." Kol turned to the avatars. "Galena, if you and the avians hover at the top of the dome, you will be in the best position to spot weapons fire. We will need you to drop down and disarm as many of the intruders as you can."

"Ren and I can create another barrier, but I do not think we can sustain it very long," Danea said. She looked tired. "A minute or two at best."

"We will hold you in reserve. Bek." Kol regarded the trainer. "Are there any weapons other than the holographite blades available to us?"

"Yes. The armory is on second level." He grimaced. "Uel and the others have probably made use of it."

"We will take whatever we can get." Kol handed out a few more assignments; then the room fell silent as we heard distant blasts echoing outside the Tåna's walls. "The fighting has already begun." He picked me up in his arms. "Go quickly and assemble your units."

As he carried me out of the room, I rested my hot face against his chest. "You're not thinking of sticking me on second level with the injured, are you?"

"I am not letting you out of my sight for as long as we live."

"Sounds like a plan." I closed my eyes for a moment as three more blasts from outside shook the walls. "Getting close."

"Yes." His stride picked up, and he carried me into third level and mounted the platform. "If we had wings, we could watch from above."

"If we had wings, I might cluck like a chicken." His puzzled frown made me grin. "Terran bird. Ill-tempered, usually ends up fried or baked."

Bek and a few of the whites came to join us, carrying a huge pallet of real blades. I saw the trainer watching me as Kol supervised issuing the weapons to the individual unit leaders.

"You knew I'd challenge Fayne, didn't you?" I asked him.

"I suspected you would." He sat down beside me. "I have never known you to accept defeat, even when it seemed inevitable."

"Why didn't you tell me Kol was still alive?"

He shook his head. "At the time, I was not certain I could get to him before they removed his memories and inserted the poison into his implant."

"Okay, I'll buy that. But why did Uel bring me here, Bek?

Why was it so important for me to become a blade dancer? He meant to keep me, like you, didn't he?"

"Yes." He averted his gaze. "As for why, you would not believe me if I told you."

So he did know. "Try me."

"Before he became what he is, he was very much like you. An outcast, driven from the only home he ever had. Uel sees himself in you." Bek moved as if to get up, but I grabbed his arm, and he sighed. "The rest you will understand when he comes for you."

I raised my brows. "Uel is that crazy?"

Bek gave me a twisted smile. "You have no idea what he's capable of, Sajora."

The surface temperature must have slowed down the Hskt-skt raiders, for control of the access passage went to a League strike force. Scouts from our welcoming committee rushed back to third level to report that nearly three hundred armed ground troops were pouring into the facility.

By then Kol had moved me to a corner and barricaded me in with Danea and Renor. I would monitor the headgear by signal and call the moves from there, while he fought with the fourth line.

"It seems I must leave you again, my heart." He ran a hand over my shoulder. "It will not be for long."

"Go stomp them into the floor." I eased into a more comfortable position against the wall. "And don't worry. Sparky will fry anyone who comes within ten feet of us."

"If I do not throttle your Chosen first," Danea muttered, still unhappy that I wouldn't go with the other injured. Then her eyes widened and she clapped a hand over them. "Your pardon, I meant no threat to your mate, warrior."

"Yes, you did." I laughed. "Oh, just relax, Sparky. He knows you're all talk." I touched Kol's arm. "Come back to me alive one more time, okay?"

His five fingers curled around my six. "I will, my heart."

We all knew we were outnumbered, and outgunned, and the

League probably assumed a certain amount of trainees would voluntarily go to either side. What they hadn't counted on was the collective fury among the trainees over discovering what the Tåna was really all about. Everyone had set aside their political differences for the moment, in favor of a little much-deserved revenge.

The rest of the welcoming committee slipped into third level, and took their positions.

"If all those who trained before us are simply slaves, then we may be the very first blade dancers to go freely from the place," Ren said to me. "What will happen when that becomes known?"

"Let's get free first; then we'll worry about finding jobs and dealing with the media." I shifted and pulled down my headgear as the first wave of League troops poured in from the corridor. Instantly all the lights went out. "Here they come." I said into my transmitter, "First team, let them get past your lines before you converge."

The strategy I'd given Kol was known as the back blitz in shockball, and just like on the field, it worked with devastating effectiveness. As the League troops moved in and struggled to get their bearings, two clusters of dancers struck from each side, coming up behind the troops to attack as fast as possible, then retreat in the opposite direction.

After the first team leader gave me the green, I called for the next. "Second team, back blitz now."

As the line of soldiers turned to the rear, two more blitz teams attacked from the front, while the first team moved in a second time from each flank.

The second team leader breathlessly transmitted that many of the League troops had now broken free of the sandwich by running through their line.

"Third team," I said. "Up and at 'em."

The fleeing soldiers were caught by the third wave, who had been lying in wait, flat on the floor in front of them. That line split in half and engulfed the soldiers, attacking for a second time from both sides. Pulse rifles fired frantically, lighting up the darkness with their exploding rounds.

"Birdie," I said. "A little help with the noise, please."

Galena and the other avatars began dropping down from the dome roof and snatching weapons, which they carried to the fourth and final team, assembled around the walls of the dome.

"Kol." I took a deep breath. "They're all yours."

As the League soldiers tried to find their way back to the level entrance, the ring of blade dancers began to move in, driving them toward an ever-decreasing center area, where the true hand-to-hand fighting took place.

I repeated the plays as more soldiers entered the dome a second time, then dispersed the teams to augment the center field. The battle drew out for what seemed to be forever to me, especially when a group of the soldiers stumbled into our barricade and I sent Danea and Renor to drive them back toward the center.

"They have left you all alone, have they?" My blades were plucked out of my hands by big furry paws. "Don't look so surprised, Jory." Thgill lifted me into his strong arms, and gave me his wolf's grin. "I said we'd run into each other again someday."

I got one solid punch to his jaw before he pinned my arm down. "You're not tinkering on me again, you Judas."

"Uel wants you back on your feet." He carried me away from the barricade and kept close to the walls, heading for the entrance to the trainee corridor. "I'm just the means."

"You pretended to be my friend so you could fuck with my leg." I spit in his face, to distract him. "I'll kill you before I let you touch me again."

"Hate me all you want. Your owner pays me to repair you, not romance you." Thgill lifted a hand to wipe his face, then jerked as I fired the pulse pistol Kol had given me into his chest. We went down on the floor. As I rolled away from him, I saw a neat, three-inch hole in his back. He choked out blood with a laugh. "Perhaps I should have romanced you." His puppy-dog eyes closed, and he went still.

Wanting more firepower, I groped for a nearby pulse rifle, still in the hands of a fallen soldier. Something tore it and the pistol from my hands and sent them skidding away; then alloy grapplers yanked me from the floor. "You can't kill me, Sajora, unless you have something that penetrates seven inches of plasteel cranium."

Funny that Bek knew he'd come back for me. "Why aren't you hiding down in the dungeon, Blade Master? We've got this under control."

He lifted me up. "Don't struggle; I'll sedate you if I must." His accumulators glowed red in the dark as he looked down at Thgill, then me. "A pity you had to kill the major. He was most convenient to have on the payroll."

"Yeah, I'm all broken up about it."

"I can see you are." He stepped over Thgill's body as he studied my face. "It still astonishes me. I never thought you would look so much like your mother."

"How would you know what she looked like?" But as soon as the words left my lips, I knew the answer.

"Because I am Samuel Kieran. Your father, Sajora."

Pulse fire shot past us from all directions, but he didn't run. We could have been taking a leisurely stroll through the park.

I tried to wriggle my good leg out of his grip, then felt the nozzle of a syrinpress against my throat and went still. "Whatever it is you want from me, you won't get it."

"You've already given it to me. Apart from your misguided penchant for honor and justice, you are everything a man could ask for in a daughter." His grapplers tightened. "Now we will leave this place, and you and I will continue as we were meant to be. Together."

"Do you have some poison for my implant?" I asked. "Because that's the only way you'll keep me with you."

"My daughter is not a slave. We will move my operation to another, safer location, and resume training. When your leg heals, I intend to place you in charge of all bladework." He dodged a soldier. "I suppose I should tell Bek he's being replaced."

I struggled against his inhuman grip. "Why haven't you killed him? And why did you save Kol from Fayne?"

"I thought keeping the Chakaran and the Jorenian alive—and under my control—will insure your obedience." He tightened his grapplers. "To keep them alive, you'll do exactly what I want."

"And what's that?"

"I made sure of that when I tested you on the *Chraeser.* You are a natural dancer, just as I was." He said that as if it were something to be proud of, like being gifted in music. "All you have to do is accept that you are your father's daughter."

I took a deep breath and shouted for Kol and help, before one of the grapplers clamped over my mouth.

"Shut up," my father said. "You don't belong with them. You're my child; you've always been mine. If your mother hadn't run from me, we would have never been apart."

A League soldier came up, pointing a weapon at me. One of Kieran's grapplers shot out and seized him by the throat. The soldier screamed, then pitched over as my father tore his head off.

I wrenched myself out of his other arm and fell, landing facedown on top of the decapitated man's body. Warm blood still spurted from his neck and sprayed in my face.

"Come along now, daughter." Kieran yanked me to my feet.

Two more soldiers came at him from either side, shoving us apart, and we both went down. One gurgled and staggered away, but the other leveled a pulse rifle at my head.

"No." Kieran rolled over on top of me, straddling me under his heavy chassis. For a split second, the light from his red eyes burned into mine. "You're mine. My daughter. Just like me."

The pulse rifle fired.

Then his plastic face began to melt, and the alloy under it glowed bright orange. He shoved me across the floor and rolled away, taking the soldier with the rifle down with him.

I covered my head with my arms, and turned away, which is why I didn't see Kieran's head explode. A shower of hot shrapnel pelted my back and neck, then a huge weight slammed into me, and I fell gratefully into the darkness.

We prevailed over the League strike force, but the battle cost us. When the last soldier fell, more than a hundred dancers littered the floor of the arena.

Kol's was the first face I saw when I regained consciousness, and I allowed myself to be a silly, emotional female and cried all over him. In between my sobs, he assured me that our clan had come through the battle unharmed, and were scattered around the Tåna, recovering the injured and helping to move the bodies of the dead.

"Sajora, there is something I must tell you now." He pressed my hand between his. "If there had been any other way, I would have seen to it."

He didn't have to say it. I'd known from the moment I'd opened my eyes. A funny rush filled my ears, and I heard myself ask, "How much is left?"

Kol helped me sit up, and drew back the linens. I took a deep breath, and then I looked.

My right leg was gone from the top of my knee down.

I'd always dreaded this moment, feared it more than anything. I'd resisted it with everything I had. Yet as I looked at the rounded, bandaged stump, I felt a sense of peace settle over me. I'd given it a good fight, all I had, and for once in my life I was going to be a good loser.

"I know I should have asked before, but"—I met his steady gaze—"how *do* you feel about chasing a one-legged woman around your bedroom?"

He put his arms around me, and I felt the wetness on his face just before he kissed me. I didn't personally start crying again until he whispered, "Impatient to see how fast you are."

Later, a long time later, Kol told me how the Hsktskt and the League had both retreated, but not because of our unexpected stand. While we were defending the Tåna, an enormous armada of ships had arrived to confront both aggressors and protect the surface from further assault.

Even more surprising, all the ships came from Joren.

"There are ships from every HouseClan under the Mother," Kol told me. "Including the HouseClans of our ClanMothers."

"Why would they come now?"

"I do not know, but there is someone here who may be able to tell you." He moved away, and I saw Enale Raska, wearing a pilot's flight suit, waiting at the foot of my berth.

"You looked better the last time we met, my ClanNiece," she said, and came closer. "I was very glad to hear that you had survived intact."

"Mostly." I glanced down at my leg. "I didn't think I'd ever hear anyone from the Raska call me Clan-anything."

"I persisted in questioning our ClanLeader until he related the entire story of your ClanMother's ordeal." She made a rather terse gesture I'd only seen Kol make when he was extremely ticked off. "We, the younger Raska, were never told. I would not have allowed you to be turned away by my ClanFather as cruelly as you were, had I known."

Maybe there was hope for the more uptight Jorenians yet. "I really didn't want to cause any trouble for your family. My mother loved your father very much." I felt suddenly old and tired and my head was starting to hurt. "I'd like to talk to you more, when I get out of here."

"I would enjoy that as well." She touched my hand. "Walk within beauty, my ClanNiece."

It took a few more days before I was healed enough to occupy a glidechair. The Jorenian surgeons who had come down to Reytalon to perform the lifesaving surgery on me felt very optimistic about my recovery.

"With the recent advances in reconstruct technology, we can fit you with a prosthesis that will return full mobility," the doctor told me. "It will also allow you to regain the sensation and reaction reflexes you had prior to amputation."

From what they'd told me, the technology was available virtually everywhere. "Why wouldn't the Tåna do that for me?"

"I questioned the healer in charge of this facility." The Jorenian male made a disgusted gesture. "This Blade Master had ordered him to do nothing but amputate. Apparently he wished to exercise some sort of control over your condition."

Or make me into what he had become. The image of Kieran's face swam in my head for a moment, but I shoved it away. My father had ended up a human brain encased in a drone body; I was simply going to have a fancy peg leg.

As soon as I was able to travel, Kol took me up to the *Wind-*

Maze, HouseClan Varena's flagship, where his ClanMother had asked the seven of us to gather for a meeting with leaders from all of the other HouseClans.

I could feel how tense Kol was as we entered the room. The 227 men and women around the huge conference table looked deadly serious.

If they were just going to banish us from Joren, no way would they send all these people to deliver the message.

There were six empty chairs, which Kol and the others occupied. I slid my glidechair into the space between Kol and Galena. For a moment everyone just looked at us.

We did have a fairly disreputable appearance. I, for obvious reasons; Kol, Nalek, and Danea were all covered with bruises and lacerations; and Galena had one of her wings bandaged. Two of Osrea's limbs were in splints, and Renor had a patch of silicon cement sprayed over one side of his head.

"The Houses of the Mother are before you." Kol's Clan-Mother, Qelta, stood up at the head of the table. "A member from every HouseClan on Joren is present here."

"How did you come to be here?" Kol asked.

"Your winged ClanSister may wish to tell you that."

Galena blushed a little as she rose to her feet. "I have been in contact with our people several times since we came to Rey-talon. My ClanMother had a transmitter implanted when I was young, in the event I ever became lost." Her iridescent eyes gleamed. "At first I just used it to reassure her that I was well. When it became clear that we would not be rescued from the Hsktskt and the League, I sent a signal to the Ruling Council, re-questing their assistance."

Well, well. I gave my little sister a grin. *Who would have thought Birdie capable of covert operations?*

"Why would you come to our aid?" Osrea wanted to know, his black tongue flickering in and out. "You have only tolerated us in the past."

"Our people became fascinated with your activities, through reports we received from Galena's ClanMother." Qelta activated a wall screen, and we watched and listened to a news broadcast from Joren. Our photoscans and details of the battle with the League and Hsktskt were reported, and schematics of the Tåna were displayed so the reporter could detail our attack strategies. "You never abandoned your Jorenian ideals, and created neutrality where none existed before. For that, you have become heroes to our people."

So we were superstars now. "You don't send eight hundred ships to get autographs," I said.

"True." Qelta smiled at me—actually *smiled* at me—then nodded to a group of Jorenians wearing ClanLeader insignia. Seven of them stood up and faced us.

One of them was Skalea, ClanLeader Raska, and he looked right at me as he spoke for the group. "We lost seven of our kin to slavers once. We could not permit it to happen again. The oath we took was one to protect you children, but in keeping our silence we formed resentment against you, who have done no wrong. We saw not the Jorenian ClanMother in each of you, but only the slavers who dishonored them. For that, your HouseClans ask your pardon."

"Why the sudden change of heart?" I couldn't help pushing. The clan needed to hear the entire truth, and not from me. "You don't treat people like dirt for twenty-five years and suddenly adore them overnight. Unless you're feeling guilty about something."

Qelta's gaze met mine. *I knew and I kept my word,* I thought. *Now tell him.*

"It is time you were told," Qelta said, accurately reading my thoughts. She looked at Kol, then took a deep breath. "I am not your biological ClanMother, Kol. Her name was Rasea Varena, and she was my ClanSister. Upon returning you safely to Joren, she put you in my charge and embraced the stars." She scanned our faces. "All of your ClanMothers embraced the stars as soon as they were reunited with their families. With the exception of Sajora's ClanMother, Kalea."

"Why?" Danea shot to her feet, her hair beginning to seethe. "Why would they commit suicide, when they had been freed?"

"Each ClanLeader suggested that the women should not live

with the shame of having borne a child out of bond," Skalea said. "Only my ClanDaughter honored you so much, Sajora, that she refused to abandon you for the stars."

It didn't give me the satisfaction I thought it would. "For which you kicked my mother off the planet."

"You knew about this, Sajora?" Kol asked me in a quiet voice.

"Yes. My mother told me everything. She also made me swear never to tell you." I looked at the six people I had grown to love. "She thought if your adopted mothers wanted to maintain that illusion, then it would be a kindness to preserve it."

"*We* could not live with their shame," Qelta said. "We discovered soon after that it was a mistake. Each HouseClan chose a replacement ClanMother, but we could not bond with you. It was not until you came, Sajora, and convinced our children to leave us that we realized how deeply we had wronged you. It drove you from us, and in your absence we saw ourselves plainly for the first time."

Kol rose to his feet, along with the others. I would have, but I contented myself with wheeling back from the table. "Is there anything more?"

"Wait, there is something I want to know." I looked at Qelta. "Do you know the names of our sires?" She nodded. "Do Kol and I share the same one?"

"No. Kol's sire was a man named Alexei Davidov. Your sire was named Samuel Kieran."

"Thanks." I took Kol's hand and squeezed it. "Told you," I murmured, just for his ears.

"We appreciate your candor." Kol took the handles of my glidechair and began pushing it toward the door.

"Kol, wait, please." Qelta came around the table, and everyone else jumped to their feet. There wasn't a happy face in the room. "The reason the Houses have assembled here is not only to tell you the truth. We wish you to return with us to Joren, to stay and allow us a second chance."

"I gathered that was your intent." Kol made a polite gesture of thanks. "I must discuss this with the members of my HouseClan. You will understand we have much to think on."

Qelta's smile wavered. "Of course. If you decide against us, we will of course escort you to any world you desire."

I waved to the rest of the room. "We'll get back to you."

I don't know why, but as we gathered in the quarters assigned to me with Kol, I thought of that first time the seven of us had come together, in the caves beneath Joren.

Jakol. Nalek. Galena. Osrea. Danea. Renor. And you, Sajora. My mother's ghost had never sounded happier. *The seven complete.*

Funny that she hadn't haunted me the whole time we were on Reytalon. A strange little pang in my heart told me I wouldn't hear from her again after this time, either.

I summoned the image of her face into my mind. *I kept my promises, Mom. And I hope I've given you a little peace. I love you; I'll never forget you. See you in the stars someday.*

I honor you, my ClanDaughter. I will be waiting.

"It would serve them well if we made them wait several rotations before returning our answer." Sparky got up and paced, until Nalek brought her a server of tea. "I spent two decades diving for scarpela pearl clusters to please my ClanMother Koralko, whom nothing pleased, and now they tell me I never shared her womb?"

"I do not mind that my ClanMother never carried me," Nalek said, scratching at the ridges on his side. "I have always imagined my birth giving her such pain."

Someone was crying softly, and I rolled over to Galena, who had her face in her hands. "It's okay, Birdie."

"It is not a smooth path." She sobbed out the words. "My ClanMother claimed everything she did was for love of me. But you were right, Sajora. I can see it now, knowing she never held me in her body. How she must hate me."

"She wouldn't have given you that transmitter if she hated you, sweetie." I would have picked her up and plopped her on my lap, but I was still on the weak side. I had to be satisfied with rubbing the top of her uninjured wing. "She honors you greatly."

"We should tell them to take us to someplace wretched and leave us there," Osrea said. "So they will forever suffer, thinking of us in such a squalid state."

"Oh, yeah, I want to go live somewhere terrible just for revenge," I said, and rolled my eyes. "Count me and Kol out of *that* plan, okay?"

It was the ever-silent Ren who finally came up with the solution. "I think we should return to Joren, but not as members of their HouseClans. We call ourselves one; we should live as one." He looked around the room. "I do not want to leave any of you. You are the one true family I have always wanted."

Everyone stared at him, then at each other.

"We wouldn't have to return to our old lives," Nalek said. "I for one would be happy to make a new start."

"Seven is rather small for a HouseClan," Kol said. "But now that we are heroes, perhaps other Jorenians would be eager to join us."

"As long as they're told I'm not Hsktskt," Os said.

"I suppose you and Kol will be ClanLeaders," Sparky said in a belligerent way.

"You can be ClanMom, if you want," I replied, then patted my Chosen's leg. "Just find your own ClanDad."

Galena stopped crying. "Will . . . will we allow other avatars, if . . . well, if others came and asked to join us?"

"Given the diversity of this bunch, sweetheart, I don't think we can turn anybody away." I wiped the tears from her face. "We'd get to be ClanSisters for real. We could stay up late and make Danea share girl talk with us."

"No girl talk, I beg you." Sparky hissed, then threw up her hands. "Oh, very well. Someone will have to look out for the six of you; it may as well be me."

"Then it's decided." I looked up at Kol. "All we need is a new name, right?"

"Yes. Names are chosen to honor what brought us together." He looked at the rest of the clan, then made a suggestion. "What say you, ClanSiblings?"

Everyone agreed it was perfect.

* * *

Joren didn't create a new HouseClan very often, and the celebration for ours drew thousands of people from all over the world. It was so huge that the Ruling Council itself hosted it, at the government pavilion in the capital of Lno.

My stump had healed enough on the journey back to Joren that when we arrived, my surgeon was able to fit me with a prosthesis, and I began the process of learning how to walk again. On the night of our celebration, I walked with my Chosen to the platform, high above the assembled Jorenians, and faced a small series of steps that looked a little like Mount Everest to me.

"If I look like I'm going to tilt over," I said to Kol, "grab me, okay?"

"I will be right beside you."

I took his hand and lifted my right leg to the first step. Slowly, with great care, I climbed the stairs. When I reached the platform, Kol put his arm around me and hugged me, and the crowd below us broke into cheers.

"This is what I get for letting everyone know I have a peg leg," I muttered.

"Smile, ClanMom," Sparky said as she passed me.

When we took our places around the ceremonial dais, the chief council member Gnelo placed beautifully woven crowns of yiborra grass on each of our heads, then turned to the silent crowd and made his speech. Then he announced the name of our HouseClan, and the serious partying began.

Several hours later, the seven of us slipped away from the celebration and borrowed some transport, and Kol drove us out to the country. He pulled off the glidepath beside a huge field of yiborra grass studded with clusters of star-shaped purple flowers.

"Look."

Above our heads, wind chasers—the Jorenian equivalent of fireflies—danced and lit up the night sky. I heard Galena laugh, and saw prisms of color shoot out from Renor's crystalline hands as he caught one of the bugs, which resembled a flying ribbon, then released it. The wind chasers seemed to like Osrea, who laughed like a boy as several lit on his head, shoulders, and arms. Danea scowled and swatted at the bugs attracted by her glow, until Nalek gently placed one on a safe spot on her insulated sleeve, and showed her its tiny wings.

Kol bent down to touch his brow to mine. "Now our journey begins, my heart."

"Walk slow for a while," I said, and touched his cheek before I looked around. "So what is this place?"

"This is the land the Ruling Council has granted to us." The others gathered around as Kol described the territorial boundaries, which included hundreds of acres of rich fields and forests, and even a small lake. "Here is where we will build our own pavilion, and begin our new lives as HouseClan Kalea."

At last, seven lonely travelers had found a home.